Funk & Wagnalls

HAMMOND

WORLD ATLAS

INCLUDING

WORLD HISTORY

SECTION

Funk & Wagnalls

DB a company of
The Dun & Bradstreet Corporation

CONTENTS

CONTENTS

GAZETTEER OF THE WORLD

This alphabetical list of grand divisions, countries, states, colonial possessions, etc., gives area, population, capital or chief town, and index references and page numbers on which they are shown on the largest scale. The index reference shows the square on the respective map in which the name of the entry may be located.

Country	Area (Sq. Miles)	Population	Capital or Chief Town	Index Ref.	Plate No.
* Afghanistan	250,775	19,280,000	Kabul	A 2	34
Africa	11,707,000	431,900,000			38-39
Alabama, U.S.A.	51,609	3,665,000	Montgomery		65
Alaska, U.S.A.	586,412	382,000	Juneau		66
* Albania	11,100	2,482,000	Tiranë	B 3	29
Alberta, Canada	255,285	1,838,037	Edmonton		58
* Algeria	919,591	16,776,000	Algiers	F 6	38
American Samoa	76	30,000	Pago Pago	J 7	41
Andorra	188	26,558	Andorra la Vella	G 1	27
* Angola	481,351	6,761,000	Luanda	K14	39
Antarctica	5,500,000				20
Antigua	171	73,000	St. Johns	G 3	45
* Argentina	1,072,070	23,983,000	Buenos Aires		43
Arizona, U.S.A.	113,909	2,270,000	Phoenix		67
Arkansas, U.S.A.	53,104	2,109,000	Little Rock		68
Ascension, St. Helena	34	1,146	Georgetown	E15	39
Asia	17,128,500	2,535,333,000			32
* Australia	2,967,909	13,684,900	Canberra		40
* Austria	32,375	7,540,000	Vienna		29
* Bahamas	5,382	197,000	Nassau	C 1	45
* Bahrain	240	300,000	Manama	F 4	33
* Bangladesh	55,126	82,900,000	Dacca	F 4	34
* Barbados	166	253,620	Bridgetown	G 4	45
* Belgium	11,781	9,813,000	Brussels		24
Belize	8,867	122,000	Belmopan	B 1	46
* Benin	43,483	3,200,000	Porto-Novo	G10	38
Bermuda	21	52,000	Hamilton	H 3	45
* Bhutan	18,147	1,200,000	Thimphu	G 3	34
* Bolivia	424,163	4,804,000	La Paz, Sucre	H 7	42
* Botswana	224,764	700,000	Gaborone	L16	39
* Brazil	3,284,426	90,840,000	Brasília		42-43
British Columbia, Canada	366,255	2,466,608	Victoria		59
British Indian Ocean Terr.	29	600	London, U.K.	L10	32
Brunei	2,226	155,000	Bandar Seri Begawan	E 5	36
* Bulgaria	42,823	8,800,000	Sofia	C 3	29
* Burma	261,789	31,240,000	Rangoon	B 2	35
* Burundi	10,747	4,100,000	Bujumbura	M12	39
California, U.S.A.	158,693	21,520,000	Sacramento		69
* Cambodia (Kampuchea)	69,898	8,110,000	Phnom Penh	D 4	35
* Cameroon	183,568	6,600,000	Yaoundé	J11	38
* Canada	3,851,809	22,992,604	Ottawa		47
* Cape Verde	1,557	302,000	Praia	H 5	19
Cayman Islands	100	10,652	Georgetown	B 3	45
* Central African Republic	236,293	1,800,000	Bangui	K10	38
Central America	197,575	19,800,000			46
* Ceylon (Sri Lanka)	25,332	14,000,000	Colombo	E 7	34
* Chad	495,752	4,178,000	N'Djamena	K 8	38
Channel Islands	74	128,000		E 6	23
* Chile	292,257	8,834,820	Santiago	G10	43
* China (People's Rep.)	3,691,000	853,000,000	Peking		37
China (Taiwan)	13,971	16,426,386	Taipei	K 7	37
* Colombia	439,513	21,117,000	Bogotá	F 3	42
Colorado, U.S.A.	104,247	2,583,000	Denver		70
* Comoros	719	266,000	Moroni	P14	39
* Congo	132,046	1,400,000	Brazzaville	J12	39
Connecticut, U.S.A.	5,009	3,117,000	Hartford		71
Cook Islands	91	17,046	Avarua	K 7	41
* Costa Rica	19,575	1,800,000	San José	C 3	46
* Cuba	44,206	8,553,395	Havana	B 2	45
* Cyprus	3,473	639,000	Nicosia	B 3	33
* Czechoslovakia	49,373	14,900,000	Prague		29
Delaware, U.S.A.	2,057	582,000	Dover		83
* Denmark	16,629	5,065,313	Copenhagen	B 3	24
District of Columbia, U.S.A.	67	702,000	Washington	B 5	83
* Djibouti	8,880	250,000	Djibouti	P 9	38
* Dominica	290	70,302	Roseau	G 4	45
* Dominican Republic	18,704	4,011,589	Santo Domingo	D 3	45
* Ecuador	109,483	6,144,000	Quito	E 4	42
* Egypt	386,659	37,900,000	Cairo	M 6	38
* El Salvador	8,260	3,418,455	San Salvador	B 2	46
England, U.K.	50,516	46,417,600	London	F 4	23
* Equatorial Guinea	10,831	320,000	Malabo	H11	39

Country	Area (Sq. Miles)	Population	Capital or Chief Town	Index Ref.	Plate No.
* Ethiopia	471,776	27,946,000	Addis Ababa	O 9	38
Europe	4,057,000	666,116,000			21
Faerøe Is., Denmark	540	38,000	Tórshavn	D 2	21
Falkland Is.	4,618	2,000	Stanley	H14	43
* Fiji	7,055	569,468	Suva	H 8	41
* Finland	130,128	4,729,000	Helsinki	E 2	24
Florida, U.S.A.	58,560	8,421,000	Tallahassee		72
* France	210,038	53,300,000	Paris		26
French Guiana	35,135	48,000	Cayenne	K 3	42
French Polynesia	1,544	135,000	Papeete	M 8	41
* Gabon	103,346	526,000	Libreville	J12	39
* Gambia	4,127	524,000	Banjul	C 9	38
Georgia, U.S.A.	58,876	4,970,000	Atlanta		73
* Germany, East (German Democratic Republic)	41,768	16,850,000	Berlin		25
* Germany, West (Federal Republic)	95,985	61,644,600	Bonn		25
* Ghana	92,099	9,900,000	Accra	F10	38
Gibraltar	2.28	30,000	Gibraltar	D 4	27
Gilbert Islands (Kiribati)	264	56,000	Bairiki	J 6	41
* Great Britain & Northern Ireland (United Kingdom)	94,399	56,076,000	London		23
* Greece	50,944	9,046,000	Athens	C 4	29
Greenland, Denmark	840,000	54,000	Nûk	B12	20
* Grenada	133	96,000	St. George's	G 4	45
Guadeloupe & Dependencies	687	324,000	Basse-Terre	F 3	45
Guam	212	111,000	Agaña	E 4	41
* Guatemala	42,042	5,200,000	Guatemala	B 2	46
* Guinea	94,925	4,500,000	Conakry	D 9	38
* Guinea-Bissau	13,948	517,000	Bissau	C 9	38
* Guyana	83,000	763,000	Georgetown	J 2	42
* Haiti	10,694	4,867,190	Port-au-Prince	D 3	45
Hawaii, U.S.A.	6,450	887,000	Honolulu		74
* Holland (Netherlands)	15,892	13,800,000	Amsterdam, The Hague		24
* Honduras	43,277	2,495,000	Tegucigalpa	C 2	46
Hong Kong	403	4,400,000	Victoria	H 7	37
* Hungary	35,919	10,590,000	Budapest		29
* Iceland	39,768	220,000	Reykjavík	C 2	21
Idaho, U.S.A.	83,557	831,000	Boise		75
Illinois, U.S.A.	56,400	11,229,000	Springfield		76
* India	1,269,339	605,614,000	New Delhi		34
Indiana, U.S.A.	36,291	5,302,000	Indianapolis		77
* Indonesia	788,430	131,255,000	Djakarta		36
Iowa, U.S.A.	56,290	2,870,000	Des Moines		78
* Iran	636,293	32,900,000	Tehran	F 3	33
* Iraq	172,476	11,400,000	Baghdad	D 3	33
* Ireland	27,136	3,109,000	Dublin		23
Isle of Man	227	59,000	Douglas	D 3	23
* Israel	7,847	3,459,000	Jerusalem		31
* Italy	116,303	56,110,000	Rome		28
* Ivory Coast	127,520	6,673,013	Abidjan	E10	38
* Jamaica	4,411	1,972,000	Kingston	C 3	45
* Japan	145,730	112,200,000	Tokyo		36
* Jordan	37,737	2,700,000	Amman		31
Kalâtdlit-Nunât (Greenland)	840,000	54,000	Nûk	B12	20
Kampuchea (Cambodia)	69,898	8,110,000	Phnom Penh	D 4	35
Kansas, U.S.A.	82,264	2,310,000	Topeka		79
Kentucky, U.S.A.	40,395	3,428,000	Frankfort		80
* Kenya	224,960	13,300,000	Nairobi	O11	39
Kiribati	264	56,000	Bairiki	J 6	41
Korea, North	46,540	17,000,000	P'yŏngyang	C 2	36
Korea, South	38,175	34,688,079	Seoul	C 3	36
* Kuwait	6,532	1,100,000	Al Kuwait	E 4	33
* Laos	91,428	3,500,000	Vientiane	E 3	35
* Lebanon	4,015	3,207,000	Beirut	C 3	33
* Lesotho	11,720	1,100,000	Maseru	M17	39
* Liberia	43,000	1,600,000	Monrovia	E10	38
* Libya	679,358	2,500,000	Tripoli	K 6	38
Liechtenstein	61	25,000	Vaduz	E 1	27
Louisiana, U.S.A.	48,523	3,841,000	Baton Rouge		81
* Luxembourg	999	358,000	Luxembourg	H 8	24

* Members of the United Nations

GAZETTEER OF THE WORLD

Country	Area (Sq. Miles)	Population	Capital or Chief Town	Index Ref.	Plate No.
Macao	6	300,000	Macao	H 7	37
Maine, U.S.A.	33,215	1,070,000	Augusta		82
*Madagascar	226,657	7,700,000	Antananarivo	R16	39
*Malawi	45,747	5,100,000	Lilongwe	N14	39
Malaya, Malaysia	50,806	9,000,000	Kuala Lumpur	D 6	35
*Malaysia	128,308	12,368,000	Kuala Lumpur	D 5	36
*Maldives	115	136,000	Male	C 8	34
*Mali	464,873	5,800,000	Bamako	E 9	38
*Malta	122	319,000	Valletta	E 7	28
Manitoba, Canada	251,000	1,021,506	Winnipeg		56
Martinique	425	332,000	Fort-de-France	G 4	45
Maryland, U.S.A.	10,577	4,144,000	Annapolis		83
Massachusetts, U.S.A.	8,257	5,809,000	Boston		84
*Mauritania	397,354	1,140,000	Nouakchott	D 8	38
*Mauritius	790	899,000	Port Louis	S19	39
Mayotte	144	40,000	Mamoutzou	P14	39
*Mexico	761,601	48,313,438	Mexico City		46
Michigan, U.S.A.	58,216	9,104,000	Lansing		85
Midway Islands	2	2,220		B 5	74
Minnesota, U.S.A.	84,068	3,965,000	St. Paul		86
Mississippi, U.S.A.	47,716	2,354,000	Jackson		87
Missouri, U.S.A.	69,686	4,778,000	Jefferson City		88
Monaco	368 acres	23,035	Monaco	G 6	26
*Mongolia	606,163	1,500,000	Ulan Bator	F 2	37
Montana, U.S.A.	147,138	753,000	Helena		89
Montserrat	40	12,300	Plymouth	G 3	45
*Morocco	172,413	16,800,000	Rabat	E 5	38
*Mozambique	308,641	9,300,000	Maputo	N16	39
Namibia	317,827	883,000	Windhoek	K16	39
Nauru	7.7	8,000	Yaren dist.	G 6	41
Nebraska, U.S.A.	77,227	1,553,000	Lincoln		90
*Nepal	54,663	12,900,000	Kathmandu	E 3	34
*Netherlands	15,892	13,800,000	Amsterdam, The Hague		24
Netherlands Antilles	390	220,000	Willemstad	E 4	45
Nevada, U.S.A.	110,540	610,000	Carson City		91
New Brunswick, Canada	28,354	677,250	Fredericton		52
New Caledonia & Dependencies	7,335	136,000	Nouméa	G 8	41
Newfoundland, Canada	156,185	557,725	St. John's		50
New Hampshire, U.S.A.	9,304	822,000	Concord		92
New Hebrides (Vanuatu)	5,700	80,000	Vila	G 7	41
New Jersey, U.S.A.	7,836	7,336,000	Trenton		93
New Mexico, U.S.A.	121,666	1,168,000	Santa Fe		94
New York, U.S.A.	49,576	18,084,000	Albany		95
*New Zealand	103,736	3,121,904	Wellington	L 7	40
*Nicaragua	45,698	1,984,000	Managua	C 2	46
*Niger	489,189	4,700,000	Niamey	H 8	38
*Nigeria	379,628	83,800,000	Lagos	H10	38
Niue	100	2,992	Alofi	K 7	41
North America	9,363,000	314,000,000			44
North Carolina, U.S.A.	52,586	5,469,000	Raleigh		96
North Dakota, U.S.A.	70,665	643,000	Bismarck		97
Northern Ireland, U.K.	5,452	1,537,200	Belfast	C 3	23
Northwest Territories, Canada	1,304,903	42,609	Yellowknife		60
*Norway	125,053	4,027,000	Oslo	B 2	24
Nova Scotia, Canada	21,425	828,571	Halifax		51
Ohio, U.S.A.	41,222	10,690,000	Columbus		98
Oklahoma, U.S.A.	69,919	2,766,000	Oklahoma City		99
*Oman	120,000	800,000	Muscat	G 5	33
Ontario, Canada	412,582	8,264,465	Toronto		54-55
Oregon, U.S.A.	96,981	2,329,000	Salem		100
Pacific Islands, Terr. of the (U.S. Trust.)	707	120,000	Kolonia	F 5	41
*Pakistan	310,403	72,370,000	Islamabad	B 3	34
*Panama	29,856	1,469,993	Panamá	D 3	46
*Papua New Guinea	183,540	2,800,000	Port Moresby	B 7	36
*Paraguay	157,047	2,314,000	Asunción	J 8	43
Pennsylvania, U.S.A.	45,333	11,862,000	Harrisburg		101
*Persia (Iran)	636,293	32,900,000	Tehran	F 3	33
*Peru	496,222	13,586,300	Lima	E 5	42
*Philippines	115,707	43,751,000	Manila	H 4	36
Pitcairn Islands	18	67	Adamstown	O 8	41
*Poland	120,725	34,364,000	Warsaw		31
*Portugal	35,549	8,825,000	Lisbon		27
Prince Edward I., Canada	2,186	118,229	Charlottetown	E 2	51
Puerto Rico	3,435	2,712,033	San Juan	G 1	45
*Qatar	4,247	150,000	Doha	F 4	33
Quebéc, Canada	594,860	6,234,445	Québec		53-54
Réunion	969	475,700	St-Denis	R20	39
Rhode Island, U.S.A.	1,214	927,000	Providence		84
Rhodesia (Zimbabwe)	150,803	6,600,000	Salisbury	M15	39
*Rumania	91,699	21,500,000	Bucharest	D 2	29
*Rwanda	10,169	4,241,000	Kigali	N12	39

Country	Area (Sq. Miles)	Population	Capital or Chief Town	Index Ref.	Plate No.
Sabah, Malaysia	28,460	633,000	Kota Kinabalu	F 4	36
St. Christopher-Nevis-Anguilla	138	56,000	Basseterre	F 3	45
St. Helena	47	4,707	Jamestown	E15	39
*St. Lucia	238	110,000	Castries	G 4	45
St-Pierre & Miquelon	93.5	6,000	St-Pierre	C 4	50
St. Vincent & the Grenadines	150	89,129	Kingstown	G 4	45
San Marino	23.4	20,000	San Marino	D 3	28
*São Tomé e Príncipe	372	80,000	São Tomé	H11	39
Sarawak, Malaysia	48,050	950,000	Kuching	E 5	36
Saskatchewan, Canada	251,700	921,323	Regina		57
*Saudi Arabia	829,995	7,200,000	Riyadh, Mecca	D 4	33
Scotland, U.K.	30,414	5,261,000	Edinburgh	D 2	23
*Senegal	75,954	5,085,388	Dakar	D 9	38
*Seychelles	145	60,000	Victoria	J10	32
*Siam (Thailand)	198,455	42,700,000	Bangkok	D 3	35
*Sierra Leone	27,925	3,100,000	Freetown	D10	38
*Singapore	226	2,300,000	Singapore	F 6	35
*Solomon Islands	11,500	196,708	Honiara	G 6	41
*Somalia	246,200	3,170,000	Mogadishu	R10	38
*South Africa	458,179	24,400,000	Cape Town, Pretoria	L18	39
South America	6,875,000	186,000,000			42-43
South Carolina, U.S.A.	31,055	2,848,000	Columbia		102
South Dakota, U.S.A.	77,047	686,000	Pierre		103
South-West Africa (Namibia)	317,827	883,000	Windhoek	K16	39
*Spain	194,881	36,000,000	Madrid		27
*Sri Lanka	25,332	14,000,000	Colombo	E 7	34
*Sudan	967,494	18,347,000	Khartoum	M 9	38
*Suriname	55,144	389,000	Paramaribo	J 3	42
*Swaziland	6,705	500,000	Mbabane	N17	39
*Sweden	173,665	8,236,461	Stockholm	C 2	24
Switzerland	15,943	6,489,000	Bern		27
*Syria	71,498	7,585,000	Damascus	C 2	33
*Tanzania	363,708	15,506,000	Dar es Salaam	N13	39
Tennessee, U.S.A.	42,244	4,214,000	Nashville		104
Texas, U.S.A.	267,338	12,487,000	Austin		105
*Thailand	198,455	42,700,000	Bangkok	D 3	35
*Togo	21,622	2,300,000	Lomé	G10	38
Tokelau	3.9	1,603	Fakaofo	J 6	41
Tonga	270	102,000	Nuku'alofa	J 8	41
*Trinidad & Tobago	1,980	1,040,000	Port of Spain	G 5	45
Tristan da Cunha	38	292	Edinburgh	J 7	19
*Tunisia	63,170	5,776,000	Tunis	H 5	38
*Turkey	300,946	40,284,000	Ankara	B 2	33
Turks & Caicos Islands	166	6,000	Cockburn Town	D 2	45
Tuvalu	10	5,887	Fongafale	H 6	41
*Uganda	91,076	11,400,000	Kampala	N11	39
*Ukrainian S.S.R., U.S.S.R.	233,089	49,438,000	Kiev	C 5	30
*Union of Soviet Socialist Republics	8,649,490	258,402,000	Moscow		30
*United Arab Emirates	32,278	240,000	Abu Dhabi	F 5	33
*United Kingdom	94,399	56,076,000	London		23
*United States of America, land land & water	3,554,609 3,615,123	214,659,000	Washington		61
*Upper Volta	105,869	6,144,013	Ouagadougou	F 9	38
*Uruguay	72,172	2,900,000	Montevideo	J10	43
Utah, U.S.A.	84,916	1,228,000	Salt Lake City		106
Vanuatu	5,700	112,596	Vila	G 7	41
Vatican City	116 acres	704		B 6	28
*Venezuela	352,143	10,398,907	Caracas	G 2	42
Vermont, U.S.A.	9,609	476,000	Montpelier		107
*Vietnam	128,405	46,600,000	Hanoi	E 3	35
Virginia, U.S.A.	40,817	5,032,000	Richmond		108
Virgin Islands, British	59	10,484	Road Town	H 1	45
Virgin Islands (U.S.A.)	133	62,468	Charlotte Amalie	G 1	45
Wake Island	2.5	437		G 4	41
Wales, U.K.	8,017	2,778,000	Cardiff	E 4	23
Wallis & Futuna	106	9,000	Matautu	J 7	41
Washington, U.S.A.	68,192	3,612,000	Olympia		109
Western Sahara	102,702	139,000		D 7	38
*Western Samoa	1,133	159,000	Apia	J 7	41
West Virginia, U.S.A.	24,181	1,821,000	Charleston		110
*White Russian (Byelorussian) S.S.R., U.S.S.R.	80,154	9,522,000	Minsk	C 4	30
Wisconsin, U.S.A.	56,154	4,609,000	Madison		111
World	57,970,000	4,240,700,000			17,19
Wyoming, U.S.A.	97,914	390,000	Cheyenne		112
*Yemen Arab Republic	77,220	5,600,000	San'a	D 6	33
*Yemen, Peoples Democratic Republic of	111,101	1,700,000	Aden	E 7	33
*Yugoslavia	98,766	21,520,000	Belgrade	B 2	29
Yukon Territory, Canada	207,076	21,836	Whitehorse		60
*Zaire	918,962	25,600,000	Kinshasa	L12	39
*Zambia	290,586	4,936,000	Lusaka	M14	39
*Zimbabwe	150,803	6,600,000	Salisbury	M15	39

GAZETTEER OF THE UNITED STATES

This section lists the major cities and all state capitals and territorial capitals of the United States. Listings for the states and territories can be found on pages 4 and 5. Population figures are derived from the 1970 U.S. Final Census, as revised.

Name	Index Ref.	Plate No.
Abilene, Tex., 89,653	E 5	105
Abington, Pa., 62,786	M 5	101
Agaña (cap.), Guam, 2,119	E 4	41
Akron, Ohio, 275,425	G 3	98
Alameda, Calif., 70,968	J 2	69
Albany, Ga., 72,623	D 7	73
Albany (cap.), N.Y., 115,781	N 5	95
Albuquerque, N. Mex., 243,751	C 3	94
Alexandria, La., 41,557	E 4	81
Alexandria, Va., 110,927	L 3	108
Alhambra, Calif., 62,125	C10	69
Allen Park, Mich., 40,747	B 7	85
Allentown, Pa., 109,871	L 4	101
Alton, Ill., 39,700	A 6	76
Altoona, Pa., 63,115	F 4	101
Amarillo, Tex.,127,010	C 2	105
Ames, Iowa, 39,505	F 4	78
Anaheim, Calif., 166,408	D11	69
Anchorage, Alaska, 49,126	B 1	66
Anderson, Ind., 70,787	F 4	77
Annapolis (cap.), Md.,30,095	H 5	83
Ann Arbor, Mich., 99,797	F 6	85
Appleton, Wis., 56,377	J 7	111
Arcadia, Calif., 45,138	C10	69
Arden-Arcade, Calif., 82,492	B 8	69
Arlington, Mass., 53,534	C 6	84
Arlington, Tex., 90,032	F 2	105
Arlington, Va., 174,284	K 3	108
Arlington Hts., Ill., 64,884	A 1	76
Arvada, Colo., 49,083	J 3	70
Asheville, N.C., 57,681	E 8	96
Athens, Ga., 44,342	F 3	73
Atlanta (cap.), Ga., 497,421	D 3	73
Atlantic City, N.J., 47,859	E 5	93
Augusta, Ga., 59,864	J 4	73
Augusta (cap.), Maine, 21,945	D 7	82
Aurora, Colo., 74,974	K 3	70
Aurora, Ill., 74,182	E 2	76
Austin (cap.), Tex., 251,808	G 7	105
Bakersfield, Calif., 69,515	G 8	69
Baldwin Park, Calif., 47,285	D10	69
Baltimore, Md., 905,787	H 3	83
Baton Rouge (cap.), La., 165,921	K 2	81
Battle Creek, Mich., 38,931	D 6	85
Bayamón, P.R., 147,552	G 1	45
Bay City, Mich., 49,449	F 5	85
Bayonne, N.J., 72,743	B 2	93
Baytown, Tex., 43,980	L 2	105
Beaumont, Tex., 117,548	K 7	105
Belleville, Ill., 41,699	B 6	76
Bellevue, Wash., 61,196	B 2	109
Bellflower, Calif., 51,454	C11	69
Bellingham, Wash., 39,375	C 2	109
Berkeley, Calif., 116,716	J 2	69
Berwyn, Ill., 52,502	B 2	76
Bethesda, Md., 71,621	A 4	83
Bethlehem, Pa., 72,686	M 4	101
Beverly, Mass., 38,348	E 5	84
Billings, Mont., 61,581	H 5	89
Biloxi, Miss., 48,486	G10	87
Binghamton, N.Y., 64,123	J 6	95
Birmingham, Ala., 300,910	D 3	65
Bismarck (cap.), N. Dak., 34,703	J 6	97
Bloomfield, N.J., 52,029	B 2	93
Bloomington, Ill., 39,992	D 3	76
Bloomington, Ind., 43,262	D 6	77
Bloomington, Minn., 81,970	G 6	86
Boise (cap.), Idaho, 74,990	B 6	75
Bossier City, La., 41,595	C 1	81
Boston (cap.), Mass., 641,071	D 7	84
Boulder, Colo., 66,870	J 2	70
Bridgeport, Conn., 156,542	C 4	71
Bristol, Conn., 55,487	D 2	71
Bristol, Pa., 67,498	N 5	101
Brockton, Mass., 89,040	K 4	84
Brookline, Mass., 58,689	C 7	84
Brownsville, Tex., 52,522	G12	105
Buena Park, Calif., 63,646	D11	69
Buffalo, N.Y., 462,768	B 5	95
Burbank, Calif., 88,871	C10	69
Burlington, Vt., 38,633	A 2	107
Caguas, P.R., 63,215	G 1	45
Cambridge, Mass.,100,361	C 6	84
Camden, N.J., 102,551	B 3	93
Canton, Ohio, 110,053	H 4	98
Carson, Calif., 71,150	C11	69
Carson City (cap.), Nev., 15,468	B 3	91
Casper, Wyo., 39,361	F 3	112
Catonsville, Md., 54,812	H 3	83
Cedar Rapids, Iowa, 110,642	K 5	78
Champaign, Ill., 56,837	E 3	76
Charleston, S.C., 66,945	G 6	102
Charleston (cap.), W.Va., 71,505	C 4	110
Charlotte, N.C., 241,178	D 4	96
Charlotte Amalie (cap.), Virgin Is., 12,220	H 1	45
Charlottesville, Va., 38,880	G 4	108
Chattanooga, Tenn., 119,923	K 4	104
Cheektowaga, N.Y., 113,844	C 5	95
Cheltenham, Pa., 40,238	M 5	101
Cherry Hill, N.J., 64,395	B 3	93
Chesapeake, Va., 89,580	M 7	108
Chester, Pa., 56,331	L 7	101
Cheyenne (cap.), Wyo., 40,914	H 4	112
Chicopee, Mass., 66,676	D 4	84
Chicago, Ill., 3,369,357	B 2	76
Chicago Hts., Ill., 40,900	B 3	76
Chula Vista, Calif., 67,901	J11	69
Cicero, Ill., 67,058	B 2	76
Cincinnati, Ohio, 451,455	B 9	98
Clearwater, Fla., 52,074	B 2	72
Clifton, N.J., 82,437	B 2	93
Cleveland, Ohio, 750,879	H 9	98
Cleveland Hts., Ohio, 60,767	H 9	98
Colorado Springs, Colo., 135,060	K 5	70
Columbia, Mo., 58,812	H 5	88
Columbia (cap.), S.C., 113,542	F 4	102
Columbus, Ga., 167,377	C 6	73
Columbus (cap.), Ohio, 540,025	E 6	98
Compton, Calif., 78,547	C11	69
Concord, Calif., 85,164	K 1	69
Concord (cap.), N.H., 30,022	C 5	92
Coral Gables, Fla., 42,494	B 5	72
Corpus Christi, Tex., 204,525	G10	105
Costa Mesa, Calif., 72,660	D11	69
Council Bluffs, Iowa, 60,348	B 6	78
Covington, Ky., 52,535	K 1	80
Cranston, R.I., 74,287	J 5	84
Cuyahoga Falls, Ohio, 49,678	G 3	98
Dallas, Tex., 844,401	H 2	105
Daly City, Calif., 66,922	H 2	69
Danbury, Conn., 50,781	B 3	71
Danville, Ill., 42,570	F 3	76
Danville, Va., 46,391	E 7	108
Davenport, Iowa, 98,469	M 5	78
Dayton, Ohio, 242,917	B 6	98
Daytona Bch., Fla., 45,327	F 2	72
Dearborn, Mich., 104,199	B 7	85
Dearborn Heights, Mich., 80,069	B 7	85
Decatur, Ala., 38,044	D 1	65
Decatur, Ill., 90,397	E 4	76
Denton, Tex., 39,874	G 4	105
Denver (cap.), Colo., 514,678	K 3	70
Des Moines (cap.), Iowa, 201,404	G 5	78
Des Plaines, Ill., 57,239	A 1	76
Detroit, Mich., 1,513,601	B 7	85
Dover (cap.), Del., 17,488	M 4	83
Downey, Calif., 88,442	C11	69
Dubuque, Iowa, 62,309	M 3	78
Duluth, Minn., 100,578	F 4	86
Dundalk, Md., 85,377	J 3	83
Durham, N.C., 95,438	H 2	96
East Chicago, Ind., 46,982	C 1	77
E. Cleveland, Ohio, 39,600	H 9	98
E. Detroit, Mich., 45,920	B 6	85
E. Hartford, Conn., 57,583	E 1	71
E. Lansing, Mich., 47,540	E 6	85
E. Los Angeles, Calif., 105,033	C10	69
E. Orange, N.J., 75,471	B 2	93
E. Point, Ga., 39,315	C 3	73
E. Providence, R.I., 48,207	J 5	84
E. St. Louis, Ill., 69,996	B 6	76
Eau Claire, Wis., 44,619	D 6	111
Edina, Minn., 44,046	G 5	86
Edison, N.J., 67,120	E 2	93
El Cajon, Calif., 52,273	J11	69
Elgin, Ill., 55,691	E 1	76
Elizabeth, N.J., 112,654	B 2	93
Elkhart, Ind., 43,152	F 1	77
Elmhurst, Ill., 48,887	A 2	76
Elmira, N.Y., 39,945	G 6	95
El Monte, Calif., 69,892	D10	69
El Paso, Tex., 322,261	A10	105
Elyria, Ohio, 53,427	F 3	98
Enfield, Conn., 46,189	E 1	71
Enid, Okla., 44,986	G 2	99
Erie, Pa., 129,231	B 1	101
Euclid, Ohio, 71,552	J 9	98
Eugene, Oreg., 79,028	D 3	100
Evanston, Ill., 80,113	B 1	76
Evansville, Ind., 138,764	C 9	77
Everett, Mass., 42,485	D 6	84
Everett, Wash., 53,622	C 3	109
Fairfield, Calif., 44,146	K 1	69
Fairfield, Conn., 56,487	B 4	71
Fall River, Mass., 96,898	K 6	84
Fargo, N.D., 53,365	S 6	97
Fayetteville, N.C., 53,510	H 4	96
Fitchburg, Mass., 43,343	G 2	84
Flint, Mich., 193,717	F 6	85
Florissant, Mo., 65,903	P 2	88
Fort Collins, Colo., 43,337	J 1	70
Ft. Lauderdale, Fla., 139,590	C 4	72
Ft. Smith, Ark., 62,802	B 3	68
Ft. Wayne, Ind., 178,021	G 2	77
Ft. Worth, Tex., 393,476	E 2	105
Framingham, Mass., 64,048	A 7	84
Frankfort (cap.), Ky., 21,902	H 4	80
Freeport, N.Y., 40,374	B 3	95
Fremont, Calif., 100,869	K 3	69
Fresno, Calif., 165,972	F 7	69
Fullerton, Calif., 85,987	D11	69
Gadsden, Ala., 53,928	G 2	65
Gainesville, Fla., 64,510	D 2	72
Galveston, Tex., 61,809	L 3	105
Gardena, Calif., 41,021	C11	69
Garden Grove, Calif., 121,357	D11	69
Garfield Hts., Ohio, 41,417	J 9	98
Garland, Tex., 81,437	H 1	105
Gary, Ind., 175,415	C 1	77
Gastonia, N.C., 47,142	C 4	96
Glendale, Calif., 132,664	C10	69
Grand Forks, N. Dak., 39,008	R 4	97
Grand Prairie, Tex., 50,904	G 2	105
Grand Rapids, Mich., 197,649	D 5	85
Granite City, Ill., 40,685	B 6	76
Great Falls, Mont., 60,091	E 3	89
Greece, N.Y., 75,136	E 4	95
Greeley, Colo., 38,902	K 2	70
Green Bay, Wis., 87,809	K 6	111
Greensboro, N.C., 144,076	F 2	96
Greenville, Miss., 39,648	B 4	87
Greenville, S.C., 61,436	C 2	102
Greenwich, Conn., 59,755	A 4	71
Groton, Conn., 38,244	G 3	71
Gulfport, Miss., 40,791	F10	87
Hamden, Conn., 49,357	D 3	71
Hamilton, Ohio, 67,865	A 7	98
Hammond, Ind., 107,885	B 1	77
Hampton, Va., 120,779	M 6	108
Harrisburg (cap.), Pa., 68,061	H 5	101
Hartford (cap.), Conn., 158,017	E 1	71
Hattiesburg, Miss., 38,277	F 8	87
Haverford, Pa., 55,132	M 6	101
Haverhill, Mass., 46,120	K 1	84
Hawthorne, Calif., 53,304	C11	69
Hayward, Calif., 93,058	K 2	69
Helena (cap.), Mont., 22,730	E 4	89
Hempstead, N.Y., 39,411	A 2	95
Hialeah, Fla., 102,452	B 4	72
High Point, N.C., 63,259	E 3	96
Hoboken, N.J., 45,380	C 2	93
Hollywood, Fla., 106,873	B 4	72
Holyoke, Mass., 50,112	D 4	84
Honolulu (cap.), Hawaii, 324,871	C 4	74
Houston, Tex., 1,232,802	J 2	105
Huntington, W.Va., 74,315	A 4	110
Huntington Beach, Calif., 115,960	C11	69
Huntsville, Ala., 139,282	E 1	65
Independence, Mo., 111,630	R 5	88
Indianapolis (cap.), Ind., 746,302	E 5	77
Inglewood, Calif., 89,985	B11	69
Inkster, Mich., 38,595	B 7	85
Iowa City, Iowa, 46,850	L 5	78
Irondequoit, N.Y., 63,675	E 4	95
Irving, Tex., 97,260	G 2	105
Irvington, N.J., 59,473	B 2	93
Jackson, Mich., 45,484	E 6	85
Jackson (cap.), Miss., 153,968	D 6	87
Jackson, Tenn., 39,996	D 3	104
Jacksonville, Fla., 528,865	E 1	72
Jamestown, N.Y., 39,795	B 6	95
Janesville, Wis., 46,426	H10	111
Jefferson City (cap.), Mo., 32,407	H 5	88
Jersey City, N.J., 260,350	B 2	93
Johnstown, Pa., 42,476	D 5	101
Joliet, Ill., 78,887	E 2	76
Joplin, Mo., 39,256	C 8	88
Juneau (cap.), Alaska, 13,556	N 1	66
Kalamazoo, Mich., 85,555	D 6	85
Kansas City, Kans., 168,213	H 2	79
Kansas City, Mo., 507,330	P 5	88
Kenosha, Wis., 78,805	M 3	111
Kettering, Ohio, 71,864	B 6	98
Knoxville, Tenn., 174,587	O 3	104
Kokomo, Ind., 44,042	E 4	77
La Crosse, Wis., 51,153	D 8	111

GAZETTEER OF THE UNITED STATES

Name	Index Ref.	Plate No.
Lafayette, Ind., 44,955	D 4	77
Lafayette, La., 68,908	F 6	81
La Habra, Calif., 41,350	D11	69
Lake Charles, La., 77,998	D 6	81
Lakeland, Fla., 41,550	D 3	72
Lakewood, Calif., 83,025	C11	69
Lakewood, Colo., 92,743	J 3	70
Lakewood, Ohio, 70,173	G 9	98
La Mesa, Calif., 39,178	H11	69
Lancaster, Pa., 57,690	K 5	101
Lansing (cap.), Mich., 131,403	E 6	85
Laredo, Tex., 69,024	E10	105
Las Vegas, Nev., 125,787	F 6	91
Lawrence, Kans., 45,698	G 3	79
Lawrence, Mass., 66,915	K 2	84
Lawton, Okla., 74,470	F 5	99
Levittown, N.Y., 65,440	B 2	95
Lewiston, Maine, 41,779	C 7	82
Lexington, Ky., 108,137	H 4	80
Lima, Ohio, 53,734	B 4	98
Lincoln (cap.), Nebr., 149,518	H 4	90
Lincoln Park, Mich., 52,984	B 7	85
Linden, N.J., 41,409	A 3	93
Little Rock (cap.), Ark., 132,483	F 4	68
Livonia, Mich., 110,109	F 6	85
Long Beach, Calif., 358,879	C11	69
Longview, Tex., 45,547	K 5	105
Lorain, Ohio, 78,185	F 3	98
Los Angeles, Calif., 2,809,813	C10	69
Louisville, Ky., 361,706	F 4	80
Lowell, Mass., 94,239	J 2	84
Lubbock, Tex., 149,101	C 4	105
Lynchburg, Va., 54,083	F 6	108
Lynn, Mass., 90,294	D 6	84
Lynwood, Calif., 43,354	C11	69
Macon, Ga., 122,423	E 5	73
Madison (cap.), Wis., 171,769	H 9	111
Madison Hts., Mich., 38,599	F 6	85
Malden, Mass., 56,127	D 6	84
Manchester, Conn., 47,994	E 1	71
Manchester, N.H., 87,754	C 6	92
Mansfield, Ohio, 55,047	F 4	98
Marion, Ind., 39,607	F 3	77
Marion, Ohio, 38,646	D 4	98
Mayagüez, P.R., 68,872	F 1	45
Medford, Mass., 64,397	C 6	84
Melbourne, Fla., 40,236	F 3	72
Memphis, Tenn., 623,530	B 4	104
Meriden, Conn., 55,959	D 2	71
Meridian, Miss., 45,083	G 6	87
Mesa, Ariz., 62,853	D 5	67
Mesquite, Tex., 55,131	H 2	105
Metairie, La., 136,477	O 4	81
Miami, Fla., 334,859	B 5	72
Miami Bch., Fla., 87,072	C 5	72
Michigan City, Ind., 39,369	C 1	77
Middletown, N.J., 54,623	E 3	93
Middletown, Ohio, 48,767	A 6	98
Midland, Tex., 59,463	C 6	105
Midwest City, Okla., 48,212	H 4	99
Milford, Conn., 50,858	C 4	71
Milwaukee, Wis., 717,372	M 1	111
Minneapolis, Minn., 434,400	G 5	86
Mobile, Ala., 190,026	B 9	65
Modesto, Calif., 61,712	D 6	69
Moline, Ill., 46,237	C 2	76
Monroe, La., 56,374	F 1	81
Montclair, N.J., 44,043	B 2	93
Montebello, Calif., 42,807	C10	69
Montgomery (cap.), Ala., 133,386	F 6	65
Monterey Park, Calif., 49,166	C10	69
Montpelier (cap.), Vt., 8,609	B 2	107
Mountain View, Calif., 54,304	K 3	69
Mount Lebanon, Pa., 39,596	B 7	101
Mt. Prospect, Ill., 45,228	A 1	76
Mt. Vernon, N.Y., 72,778	H 1	95
Muncie, Ind., 69,082	G 4	77
Muskegon, Mich., 44,631	C 5	85
Nashua, N.H., 55,820	C 6	92
Nashville (cap.), Tenn., 447,877	H 2	104
National City, Calif., 43,184	J11	69

Name	Index Ref.	Plate No.
New Albany, Ind., 38,402	F 8	77
Newark, N.J., 381,930	B 2	93
Newark, Ohio, 41,836	F 5	98
New Bedford, Mass., 101,777	K 6	84
New Britain, Conn., 83,441	E 2	71
New Brunswick, N.J., 41,885	E 3	93
New Castle, Pa., 38,559	B 3	101
New Haven, Conn., 137,707	D 3	71
New Orleans, La., 593,471	O 4	81
Newport Bch., Calif., 49,422	D11	69
Newport News, Va., 138,177	L 6	108
New Rochelle, N.Y., 75,385	J 1	95
Newton, Mass., 91,263	C 7	84
New York, N.Y., 7,895,563	C 2	95
Niagara Falls, N.Y., 85,615	C 4	95
Norfolk, Va., 307,951	M 7	108
Norman, Okla., 52,117	H 4	99
Norristown, Pa., 38,169	M 5	101
North Bergen, N.J., 47,751	B 2	93
N. Charleston, S.C., 53,617	G 6	102
N. Chicago, Ill., 47,275	F 1	76
N. Little Rock, Ark., 60,040	F 4	68
Norwalk, Calif., 91,827	C11	69
Norwalk, Conn., 79,288	B 4	71
Norwich, Conn., 41,739	G 2	71
Oakland, Calif., 361,561	J 2	69
Oak Lawn, Ill., 60,305	B 2	76
Oak Park, Ill., 62,511	B 2	76
Oceanside, Calif., 40,494	H10	69
Odessa, Tex., 78,380	B 6	105
Ogden, Utah, 69,478	C 2	106
Oklahoma City (cap.), Okla., 368,377	G 4	99
Old Bridge, N.J., 48,715	E 3	93
Olympia (cap.), Wash., 23,296	C 3	109
Omaha, Nebr., 346,929	J 3	90
Ontario, Calif., 64,118	D10	69
Orange, Calif., 77,365	D11	69
Orlando, Fla., 99,006	E 3	72
Oshkosh, Wis., 53,082	J 8	111
Overland Park, Kans., 79,034	H 3	79
Owensboro, Ky., 50,329	C 5	80
Oxnard, Calif., 71,225	F 9	69
Pago Pago (cap.), American Samoa, 2,451	J 7	41
Palo Alto, Calif., 55,835	K 3	69
Parkersburg, W.Va., 44,208	D 2	110
Park Ridge, Ill., 42,614	A 1	76
Parma, Ohio, 100,216	H 9	98
Parsippany-Troy Hills, N.J., 55,112	E 2	93
Pasadena, Calif., 112,951	C10	69
Pasadena, Tex., 89,277	J 2	105
Passaic, N.J., 55,124	B 2	93
Paterson, N.J., 144,824	B 2	93
Pawtucket, R.I., 76,984	J 5	84
Peabody, Mass., 48,080	E 5	84
Pensacola, Fla., 59,507	B 6	72
Peoria, Ill., 126,963	D 3	76
Perth Amboy, N.J., 38,798	E 2	93
Petersburg, Va., 44,124	J 6	108
Philadelphia, Pa., 1,949,996	M 6	101
Phoenix (cap.), Ariz., 582,500	C 5	67
Pico Rivera, Calif., 54,170	C10	69
Pierre (cap.), S. Dak., 9,699	J 5	103
Pine Bluff, Ark., 57,389	F 5	68
Pittsburgh, Pa., 520,117	B 7	101
Pittsfield, Mass., 57,020	A 3	84
Plainfield, N.J., 46,862	E 2	93
Pocatello, Idaho, 40,036	F 7	75
Pomona, Calif., 87,384	D10	69
Pompano Bch., Fla., 38,587	F 5	72
Ponce, P.R., 128,233	G 1	45
Pontiac, Mich., 85,279	F 6	85
Port Arthur, Tex., 57,371	K 8	105
Portland, Maine, 65,116	C 8	82
Portland, Oreg., 379,967	B 2	100
Portsmouth, Va., 110,963	L 7	108
Prichard, Ala., 41,578	B 9	65
Providence (cap.), R.I., 179,116	J 5	84
Provo, Utah, 53,131	C 3	106
Pueblo, Colo., 97,453	K 6	70
Quincy, Ill., 45,288	B 4	76
Quincy, Mass., 87,966	D 7	84

Name	Index Ref.	Plate No.
Racine, Wis., 95,162	M 3	111
Raleigh (cap.), N.C., 123,793	H 3	96
Rapid City, S. Dak., 43,836	C 5	103
Reading, Pa., 87,643	L 5	101
Redondo Bch., Calif., 57,451	B11	69
Redwood City, Calif., 55,686	J 3	69
Reno, Nev., 72,863	B 3	91
Revere, Mass., 43,159	D 6	84
Richardson, Tex., 48,582	G 1	105
Richfield, Minn., 47,231	G 6	86
Richmond, Calif., 79,043	J 1	69
Richmond, Ind., 43,999	H 5	77
Richmond (cap.), Va., 249,431	K 5	108
Riverside, Calif., 140,089	E10	69
Roanoke, Va., 92,115	D 6	108
Rochester, Minn., 53,766	F 6	86
Rochester, N.Y., 296,233	E 4	95
Rockford, Ill., 147,370	D 1	76
Rock Island, Ill., 50,166	C 2	76
Rockville, Md., 41,821	F 4	83
Rome, N.Y., 50,148	J 4	95
Rosemead, Calif., 40,972	D10	69
Roseville, Mich., 60,529	G 6	85
Royal Oak, Mich., 86,238	B 6	85
Sacramento (cap.), Calif., 257,105	B 8	69
Saginaw, Mich., 91,849	F 5	85
Saint Clair Shores, Mich., 88,093	G 6	85
St. Cloud, Minn., 39,691	D 5	86
St. Joseph, Mo., 72,691	C 3	88
St. Louis, Mo., 622,236	P 3	88
St. Louis Park, Minn., 48,922	G 5	86
St. Paul (cap.), Minn., 309,714	G 5	86
St. Petersburg, Fla., 216,159	B 3	72
Salem, Mass., 40,556	E 5	84
Salem (cap.), Oreg., 68,480	A 3	100
Salinas, Calif., 58,896	D 7	69
Salt Lake City (cap.), Utah, 175,885	C 3	106
San Angelo, Tex., 63,884	D 6	105
San Antonio, Tex., 654,153	F 8	105
San Bernardino, Calif., 106,869	E10	69
San Diego, Calif., 697,027	H11	69
San Francisco, Calif., 715,674	H 2	69
San Jose, Calif., 445,779	L 3	69
San Juan (cap.), P.R., 452,749	G 1	45
San Leandro, Calif., 68,698	J 2	69
San Mateo, Calif., 78,991	J 3	69
San Rafael, Calif., 38,977	J 1	69
Santa Ana, Calif., 155,762	D11	69
Sta. Barbara, Calif., 70,215	F 9	69
Sta. Clara, Calif., 87,717	K 3	69
Sta. Fe (cap.), N. Mex., 41,167	C 3	94
Sta. Monica, Calif., 88,289	B10	69
Sta. Rosa, Calif., 50,006	C 5	69
Sarasota, Fla., 40,237	D 4	72
Savannah, Ga., 118,349	L 6	73
Schenectady, N.Y., 77,958	M 5	95
Scottsdale, Ariz., 67,823	D 5	67
Scranton, Pa., 103,564	L 3	101
Seattle, Wash., 530,831	A 2	109
Sheboygan, Wis., 48,484	L 8	111
Shreveport, La., 182,064	C 1	81
Silver Spring, Md., 77,411	B 4	83
Simi Valley, Calif., 59,832	G 9	69
Sioux City, Iowa, 85,925	A 3	78
Sioux Falls, S. Dak., 72,488	R 6	103
Skokie, Ill., 68,322	B 1	76
Somerville, Mass., 88,779	C 6	84
South Bend, Ind., 125,580	E 1	77
Southfield, Mich., 69,285	F 6	85
South Gate, Calif., 56,909	C11	69
South San Francisco, Calif., 46,646	J 2	69
Spartanburg, S.C., 44,546	C 1	102
Spokane, Wash., 170,516	H 3	109
Springfield (cap.), Ill., 91,753	D 4	76
Springfield, Mass., 163,905	D 4	84
Springfield, Mo., 120,096	F 8	88
Springfield, Ohio, 81,941	C 6	98
Stamford, Conn., 108,798	A 4	71
Sterling Heights, Mich., 61,365	B 6	85

Name	Index Ref.	Plate No.
Stockton, Calif., 109,963	D 6	69
Stratford, Conn., 49,775	C 4	71
Suffolk, Va., 45,024	L 7	108
Sunnyvale, Calif., 95,408	K 3	69
Syracuse, N.Y., 197,297	H 4	95
Tacoma, Wash., 154,407	C 3	109
Tallahassee (cap.), Fla., 72,624	B 1	72
Tampa, Fla., 277,753	C 2	72
Taunton, Mass., 43,756	K 5	84
Taylor, Mich., 70,020	B 7	85
Teaneck, N.J., 42,355	B 2	93
Tempe, Ariz., 63,550	D 5	67
Terre Haute, Ind., 70,335	C 6	77
Texas City, Tex., 38,908	K 3	105
Toledo, Ohio, 383,105	D 2	98
Topeka (cap.), Kans., 125,011	G 2	79
Torrance, Calif., 134,968	C11	69
Towson, Md., 77,768	H 3	83
Trenton (cap.), N.J., 104,786	D 3	93
Troy, Mich., 39,419	B 6	85
Troy, N.Y., 62,918	N 5	95
Tucson, Ariz., 262,933	D 6	67
Tulsa, Okla., 330,350	K 2	99
Tuscaloosa, Ala., 65,773	C 4	65
Tyler, Tex., 57,770	J 5	105
Union, N.J., 53,077	A 2	93
Union City, N.J., 57,305	B 2	93
University City, Mo., 47,527	P 3	88
Upper Arlington, Ohio, 38,727	D 6	98
Upper Darby, Pa., 95,910	M 6	101
Utica, N.Y., 91,340	K 4	95
Vallejo, Calif., 71,710	J 1	69
Valley Stream, N.Y., 40,413	A 2	95
Vancouver, Wash., 41,859	C 5	109
Ventura, Calif., 57,964	F 9	69
Victoria, Tex., 41,349	H 9	105
Vineland, N.J., 47,399	C 5	93
Virginia Bch., Va., 172,106	M 7	108
Waco, Tex., 95,326	G 6	105
Walnut Creek, Calif., 39,844	K 2	69
Waltham, Mass., 61,582	B 6	84
Warren, Mich., 179,260	B 6	85
Warren, Ohio, 63,494	J 3	98
Warwick, R.I., 83,694	J 6	84
Washington, D.C. (cap.), U.S., 756,510	B 5	83
Waterbury, Conn., 108,033	C 2	71
Waterloo, Iowa, 75,533	J 4	78
Watertown, Mass., 39,307	C 6	84
Waukegan, Ill., 65,134	F 1	76
Waukesha, Wis., 39,695	K 1	111
Wauwatosa, Wis., 58,676	L 1	111
Wayne, N.J., 49,141	A 1	93
West Allis, Wis., 71,649	L 1	111
W. Covina, Calif., 68,034	D10	69
W. Hartford, Conn., 68,031	D 1	71
W. Haven, Conn., 52,851	D 3	71
Westland, Mich., 86,749	F 6	85
Westminster, Calif., 59,874	D11	69
West New York, N.J., 40,627	C 2	93
W. Orange, N.J., 43,715	A 2	93
W. Palm Bch., Fla., 57,375	F 5	72
W. Seneca, N.Y., 48,404	C 5	95
Weymouth, Mass., 54,610	D 8	84
Wheaton, Md., 66,280	A 3	83
Wheeling, W. Va., 48,188	K 5	110
White Plains, N.Y., 50,346	J 1	95
Whittier, Calif., 72,863	D11	69
Wichita, Kans., 276,554	E 4	79
Wichita Falls, Tex., 96,265	F 4	105
Wilkes-Barre, Pa., 58,856	L 3	101
Willingboro, N.J., 43,386	D 3	93
Wilmington, Del., 80,386	M 2	83
Wilmington, N.C., 46,169	J 6	96
Winston-Salem, N.C., 133,683	E 2	96
Woodbridge, N.J., 98,944	E 2	93
Woonsocket, R.I., 46,820	J 4	84
Worcester, Mass., 176,572	H 3	84
Wyandotte, Mich., 41,061	B 7	85
Wyoming, Mich., 56,560	D 6	85
Yakima, Wash., 45,588	E 4	109
Yonkers, N.Y., 204,297	H 1	95
York, Pa., 50,335	J 6	101
Youngstown, Ohio, 140,909	J 3	98

8

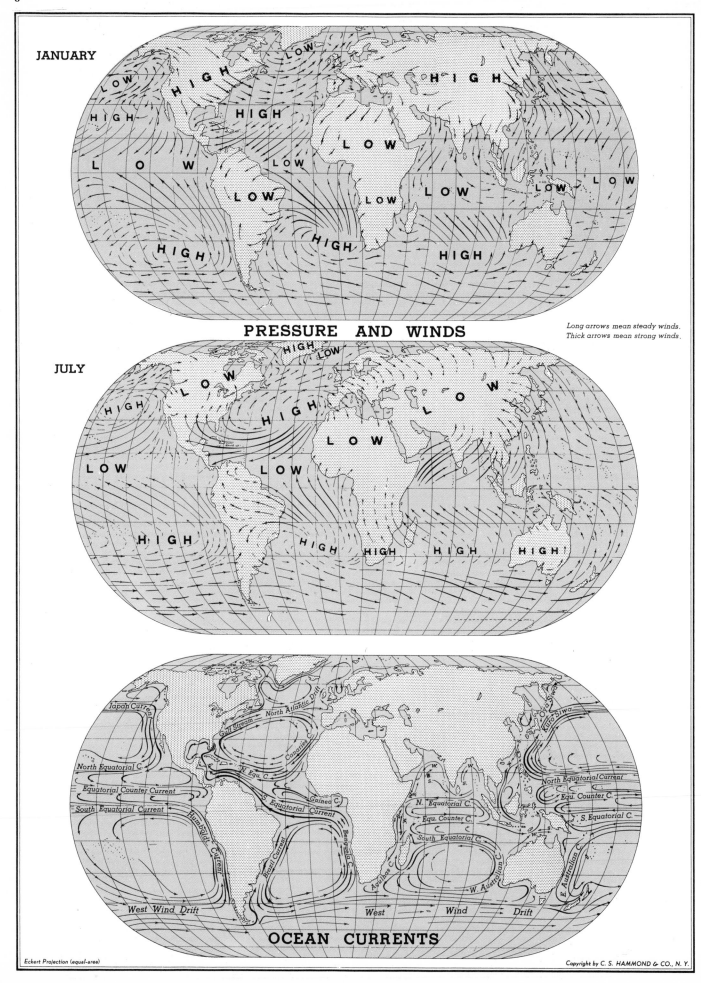

JANUARY

LOW HIGH LOW
HIGH HIGH
LOW LOW
LOW LOW LOW LOW LOW
HIGH HIGH HIGH

PRESSURE AND WINDS

Long arrows mean steady winds.
Thick arrows mean strong winds.

JULY

HIGH LOW
LOW HIGH LOW
HIGH LOW
LOW LOW
HIGH HIGH HIGH HIGH HIGH

Japan Current
North Atlantic Drift
Gulf Stream
Canaries C.
Kuro Siwo
Oya Siwo
North Equatorial C.
N. Equ. C.
Equatorial Counter Current
Guinea C.
North Equatorial Current
South Equatorial Current
S. Equatorial Current
Equ. Counter C.
N. Equatorial C.
S. Equatorial C.
Humboldt Current
Brazil Current
Equ. Counter C.
Benguela C.
South Equatorial C.
Agulhas C.
W. Australian C.
E. Australian C.
West Wind Drift
West Wind Drift

OCEAN CURRENTS

Eckert Projection (equal-area)

Copyright by C. S. HAMMOND & CO., N. Y.

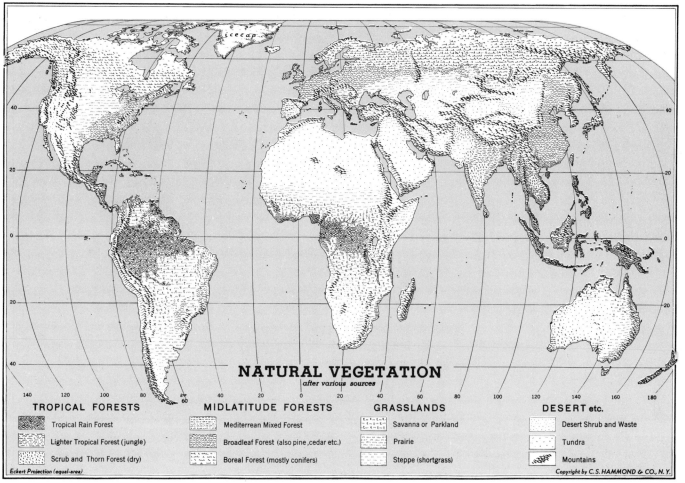

NATURAL VEGETATION
after various sources

TROPICAL FORESTS

- Tropical Rain Forest
- Lighter Tropical Forest (jungle)
- Scrub and Thorn Forest (dry)

MIDLATITUDE FORESTS

- Mediterrean Mixed Forest
- Broadleaf Forest (also pine, cedar etc.)
- Boreal Forest (mostly conifers)

GRASSLANDS

- Savanna or Parkland
- Prairie
- Steppe (shortgrass)

DESERT etc.

- Desert Shrub and Waste
- Tundra
- Mountains

Eckert Projection (equal-area)

Copyright by C. S. HAMMOND & CO., N. Y.

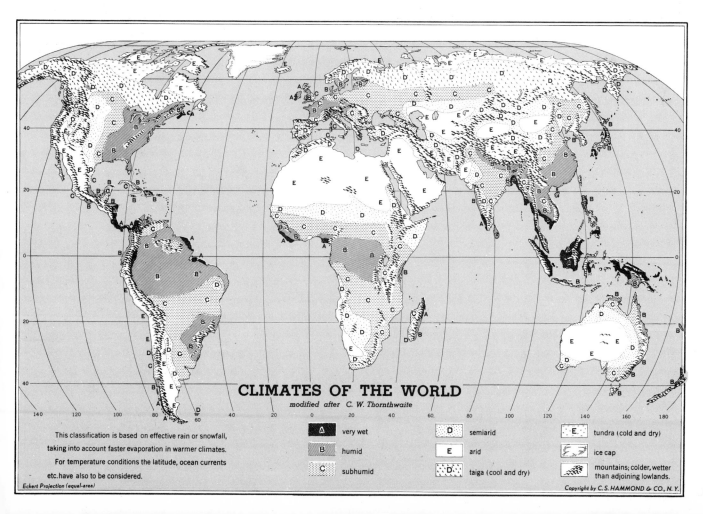

CLIMATES OF THE WORLD
modified after C. W. Thornthwaite

This classification is based on effective rain or snowfall,
taking into account faster evaporation in warmer climates.
For temperature conditions the latitude, ocean currents
etc. have also to be considered.

- **A** very wet
- **B** humid
- **C** subhumid
- **D** semiarid
- **E** arid
- **D** taiga (cool and dry)
- **E** tundra (cold and dry)
- ice cap
- mountains; colder, wetter than adjoining lowlands.

Eckert Projection (equal-area)

Copyright by C. S. HAMMOND & CO., N. Y.

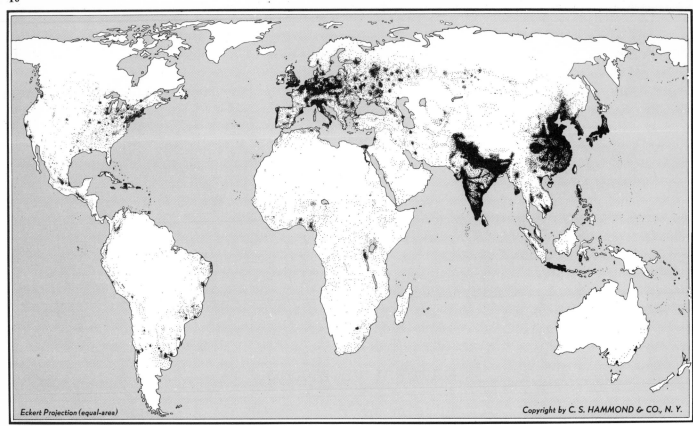

Eckert Projection (equal-area)

Copyright by C. S. HAMMOND & CO., N. Y.

DENSITY OF POPULATION. *One of the most outstanding facts of human geography is the extremely uneven distribution of people over the Earth. One-half of the Earth's surface has less than 3 people per square mile, while in the lowlands of India, China, Java and Japan rural density reaches the incredible congestion of 2000-3000 per square mile. Three-fourths of the Earth's population live in four relatively small areas; Northeastern United States, North-Central Europe, India and the Far East.*

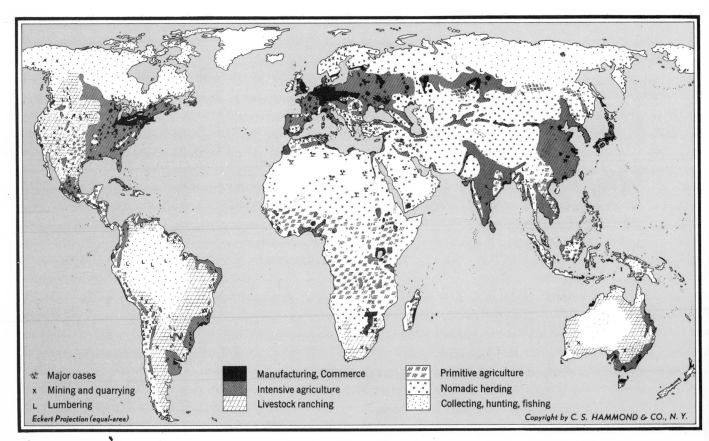

⚓ Major oases	◼ Manufacturing, Commerce	▨ Primitive agriculture
x Mining and quarrying	Intensive agriculture	Nomadic herding
L Lumbering	Livestock ranching	Collecting, hunting, fishing

Eckert Projection (equal-area)

Copyright by C. S. HAMMOND & CO., N. Y.

OCCUPATIONS. *Correlation with the density of population shows that the most densely populated areas fall into the regions of manufacturing and intensive farming. All other economies require considerable space. The most sparsely inhabited areas are those of collecting, hunting and fishing. Areas with practically no habitation are left blank.*

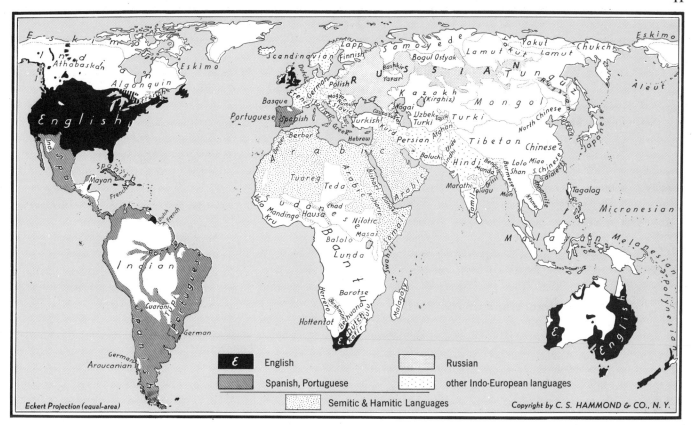

LANGUAGES. *Several hundred different languages are spoken in the World, and in many places two or more languages are spoken, sometimes by the same people. The map above shows the dominant languages in each locality. English, French, Spanish, Russian, Arabic and Swahili are spoken by many people as a second language for commerce or travel.*

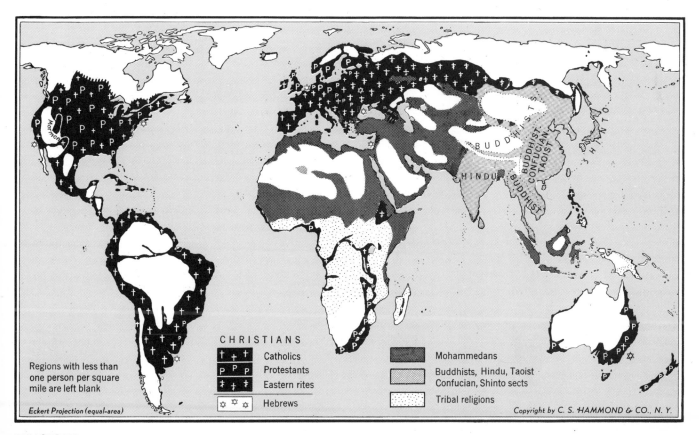

RELIGIONS. *Most people of the Earth belong to four major religions: Christians, Mohammedans, Brahmans, Buddhists and derivatives. The Eastern rites of the Christians include the Greek Orthodox, Greek Catholic, Armenian, Syrian, Coptic and more minor churches. The lamaism of Tibet and Mongolia differs a great deal from Buddhism in Burma and Thailand. In the religion of China the teachings of Buddha, Confucius and Tao are mixed, while in Shinto a great deal of ancestor and emperor worship is added. About 11 million Hebrews live scattered over the globe, chiefly in cities and in the state of Israel.*

WORLD STATISTICS

EARTH AND SOLAR SYSTEM

Elements of the Solar System

	Mean Distance From Sun in Miles	Period of Revolution Around Sun	Period of Rotation on Axis	Equatorial Diameter in Miles	Surface Gravity (Earth=1)	Mean Density (Water=1)	Number of Satellites
SUN	....	....	25.4 days	864,000	27.95	1.4	....
MERCURY	36,001,000	87.97 days	59 days	3,012	0.38	5.5	0
VENUS	67,235,000	224.70 days	247 days	7,530	0.88	5.25	0
EARTH	93,003,000	365.26 days	23h 56m	7,927	1.00	5.5	1
MARS	141,708,000	687 days	24h 37m	4,200	0.39	4.0	2
JUPITER	483,880,000	11.86 years	9h 50m	88,771	2.65	1.3	15
SATURN	887,141,000	29.46 years	10h 14m	74,187	1.17	0.7	10
URANUS	1,783,000,000	84.02 years	10h 45m	32,200	1.05	1.3	5
NEPTUNE	2,795,000,000	164.79 years	15h 48m	30,800	1.23	1.6	2
PLUTO	3,667,000,000	248.5 years	6.4 days	3,725	0.7	4.5	0

Dimensions of the Earth

Superficial area	192,251,000	sq. miles
Land surface	52,970,000	" "
North America	9,363,000	" "
South America	6,885,700	" "
Europe	4,057,000	" "
Asia	17,128,500	" "
Africa	11,707,000	" "
Australia	2,941,500	" "
Water surface	139,781,000	" "
Atlantic Ocean	31,862,000	" "
Pacific Ocean	64,186,000	" "
Indian Ocean	28,350,000	" "
Arctic Ocean	5,427,000	" "
Equatorial circumference	24,894	miles
Meridional circumference	24,811	"
Equatorial diameter	7,926.677	"
Polar diameter	7,899.980	"
Equatorial radius	3,963.34	"
Polar radius	3,949.99	"
Volume of the Earth	260,000,000,000	cubic miles
Mass, or weight	5,890,000,000,000,000,000,000	tons
Mean distance from the Sun	93,003,000	miles

The Moon is the Earth's natural satellite. The mean distance which separates the Earth from the Moon is 237,087 miles. The Moon's true period of revolution (sidereal month) is 27⅓ days. The Moon rotates on its own axis once during this time. The phase period or time between new moons (synodic month) is 29½ days. The Moon's diameter is 2,160 miles, its density is 3.3 and its surface gravity is 0.2.

PRINCIPAL LAKES AND INLAND SEAS

	AREA IN SQ. MILES
Caspian Sea	143,243
Lake Superior	31,700
Lake Victoria	26,724
Aral Sea	25,676
Lake Huron	23,010
Lake Michigan	22,300
Lake Tanganyika	12,650
Lake Baykal	12,162
Great Bear Lake	12,096
Lake Nyasa	11,555
Great Slave Lake	11,269
Lake Erie	9,910
Lake Winnipeg	9,417
Lake Ontario	7,340
Lake Ladoga	7,104
Lake Balkhash	7,027
Lake Chad	5,300
Lake Onega	3,710
Lake Titicaca	3,200
Lake Nicaragua	3,100
Lake Athabasca	3,064
Reindeer Lake	2,568
Lake Turkana (Rudolf)	2,463
Issyk-Kul'	2,425
Vanern	2,156
Lake Winnipegosis	2,075
Lake Mobutu Sese Seko	2,075
Kariba Lake	2,050
Lake Urmia	1,815
Lake of the Woods	1,679
Lake Peipus	1,400
Lake Tana	1,219
Great Salt Lake	1,100
Lake Iliamna	1,000
Vattern	733
Dead Sea	400
Lake Balaton	228
Lake Geneva	224
Lake of Constance	208
Lake Tahoe	193
Lake Garda	143
Lake Como	56
Lake of Lucerne	44
Lake of Zurich	34

OCEANS AND SEAS OF THE WORLD

	AREA IN SQ. MILES	GREATEST DEPTH IN FEET	VOLUME IN CUBIC MILES
Pacific Ocean	64,186,000	36,198	167,025,000
Atlantic Ocean	31,862,000	28,374	77,580,000
Indian Ocean	28,350,000	25,344	68,213,000
Arctic Ocean	5,427,000	17,880	3,026,000
Caribbean Sea	970,000	24,720	2,298,400
Mediterranean Sea	969,000	16,896	1,019,400
South China Sea	895,000	15,000	....
Bering Sea	875,000	15,800	788,500
Gulf of Mexico	600,000	12,300	....
Sea of Okhotsk	590,000	11,070	454,700
East China Sea	482,000	9,500	52,700
Japan Sea	389,000	12,280	383,200
Hudson Bay	317,500	846	37,590
North Sea	222,000	2,200	12,890
Black Sea	185,000	7,365	....
Red Sea	169,000	7,200	53,700
Baltic Sea	163,000	1,506	5,360

GREAT SHIP CANALS

	LENGTH IN MILES	MIN. DEPTH IN FEET
Volga-Baltic, U.S.S.R.	225	....
Baltic-White Sea, U.S.S.R.	140	16
Suez, Egypt	100.76	42
Albert, Belgium	80	16.5
Moscow-Volga, U.S.S.R.	80	18
Volgo-Don, U.S.S.R.	62	....
Gota, Sweden	54	10
Kiel, West Germany	53.2	38
Panama Canal, Panama	50.72	41.6
Houston Ship, U.S.A.	50	36
Amsterdam-Rhine, Netherlands	45	41
Beaumont-Port Arthur, U.S.A.	40	32
Manchester Ship, England	35.5	28
Chicago Sanitary and Ship, U.S.A.	33.8	20
Welland, Canada	27.6	27
Juliana, Netherlands	21	11.8
Chesapeake and Delaware, U.S.A.	19	35
Cape Cod, U.S.A.	17.4	32
Lake Washington, U.S.A.	8	30
Corinth, Greece	3.5	26.25
Sault Ste. Marie, U.S.A.	1.8	27
Sault Ste. Marie, Canada	1.4	27

WORLD STATISTICS

PRINCIPAL ISLANDS OF THE WORLD

	AREA IN SQ. MILES		AREA IN SQ. MILES		AREA IN SQ. MILES		AREA IN SQ. MILES
Greenland	840,000	Tierra del Fuego	17,900	Hebrides	2,812	Pemba	380
New Guinea	305,000	Melville	16,274	Canary Islands	2,808	Orkney Islands	372
Borneo	290,000	Southampton	15,913	Kerguelen	2,700	Madeira Islands	307
Madagascar	226,400	Solomon Islands	15,600	Prince Edward	2,170	Dominica	290
Baffin	195,928	New Britain	14,100	Trinidad and Tobago	1,980	Tonga	270
Sumatra	164,000	Taiwan (Formosa)	13,836	Balearic Islands	1,936	Molokai	261
Philippines	115,707	Kyushu	13,770	Ryukyu Islands	1,767	St. Lucia	238
New Zealand	103,736	Hainan	13,127	Madura	1,752	Corfu	229
Great Britain	88,764	Prince of Wales	12,872	Cape Verde Islands	1,557	Bornholm	227
Honshu	88,000	Vancouver	12,079	South Georgia	1,450	Isle of Man	227
Victoria	83,896	Timor	11,527	Long I., New York	1,401	Singapore	226
Ellesmere	75,767	Sicily	9,926	Socotra	1,400	Guam	212
Celebes	72,986	Somerset	9,570	Samoa	1,209	Isle Royale	196
Java	48,842	Sardinia	9,301	Gotland	1,153	Virgin Islands	192
Newfoundland	42,031	Fiji Islands	7,055	Reunion	969	Curacao	182
Cuba	40,533	Shikoku	6,860	Azores	902	Barbados	166
Luzon	40,420	New Caledonia	6,530	Isle of Pines	849	Isle of Wight	145
Iceland	39,768	Kuril Islands	6,025	Macıas Nguema Biyogo	779	Lanai	140
Mindanao	36,537	New Hebrides	5,700	Tenerife	745	St. Vincent	131
Molucca Islands	32,307	Bahama Islands	5,382	Maui	729	Maltese Islands	122
Novaya Zemlya	31,900	Falkland Islands	4,618	Mauritius	720	Grenada	120
Ireland	31,743	Jamaica	4,232	Zanzibar	641	Tobago	116
Sakhalin	29,500	Hawaii	4,038	Oahu	608	Martha's Vineyard	93
Hispaniola	29,399	Cape Breton	3,981	Guadeloupe	584	Seychelles	85
Hokkaido	28,983	Cyprus	3,572	Aland Is.	581	Channel Islands	74
Banks	27,038	Puerto Rico	3,435	Kauai	553	St. Helena	47
Tasmania	26,383	Corsica	3,352	Shetland Islands	552	Nantucket	46
Ceylon	25,332	New Ireland	3,340	Rhodes	542	Ascension	34
Svalbard	23,957	Crete	3,218	Caroline Islands	463	Hong Kong	30
Devon	21,331	Galapagos Islands	3,075	Martinique	425	Manhattan, New York	22
Bismarck Arch.	18,976	Wrangel	2,819	Tahiti	402	Bermuda Islands	21

PRINCIPAL MOUNTAINS OF THE WORLD

	FEET		FEET
Everest, Nepal-China	29,028	Kenya, Kenya	17,058
K2 (Godwin Austen), India	28,250	Ararat, Turkey	16,946
Kanchenjunga, Nepal-India	28,208	Vinson Massif, Antarctica	16,864
Lhotse, Nepal-China	27,923	Margherita (Ruwenzori), Africa	16,795
Makalu, Nepal-China	27,824	Kazbek, U.S.S.R.	16,512
Dhaulagiri, Nepal	26,810	Djaja, Indonesia	16,503
Nanga Parbat, India	26,660	Blanc, France	15,771
Annapurna, Nepal	26,504	Klyuchevskaya Sopka, U.S.S.R.	15,584
Nanda Devi, India	25,645	Rosa (Dufourspitze), Italy-	
Kamet, India	25,447	Switzerland	15,203
Tirich Mir, Pakistan	25,230	Ras Dashan, Ethiopia	15,157
Minya Konka, China	24,902	Matterhorn, Switzerland	14,688
Muztagh Ata, China	24,757	Whitney, California	14,494
Communism Peak, U.S.S.R.	24,599	Elbert, Colorado	14,433
Pobeda Peak, U.S.S.R.	24,406	Rainier, Washington	14,410
Chomo Lhari, Bhutan-China	23,997	Blanca Peak, Colorado	14,345
Muztagh, China	23,891	Markham, Antarctica	14,272
Aconcagua, Argentina	22,831	Shasta, California	14,162
Ojos del Salado, Chile-Arg.	22,572	Pikes Peak, Colorado	14,110
Tupungato, Chile-Argentina	22,310	Finsteraarhorn, Switzerland	14,022
Mercedario, Argentina	22,211	Tajumulco, Guatemala	13,845
Huascaran, Peru	22,205	Mauna Kea, Hawaii	13,796
Llullaillaco, Chile-Arg.	22,057	Mauna Loa, Hawaii	13,680
Ancohuma, Bolivia	21,489	Toubkal, Morocco	13,665
Illampu, Bolivia	21,276	Jungfrau, Switzerland	13,642
Chimborazo, Ecuador	20,561	Cameroon, Cameroon	13,350
McKinley, Alaska	20,320	Gran Paradiso, Italy	13,323
Logan, Yukon	19,524	Robson, British Columbia	12,972
Cotopaxi, Ecuador	19,347	Grossglockner, Austria	12,461
Kilimanjaro, Tanzania	19,340	Fuji, Japan	12,389
El Misti, Peru	19,101	Cook, New Zealand	12,349
Huila, Colombia	18,865	Pico de Teide, Canary Is.	12,172
Citlaltepetl (Orizaba), Mexico	18,855	Semeru, Java, Indonesia	12,060
El'brus, U.S.S.R.	18,510	Mulhacen, Spain	11,411
Demavend, Iran	18,376	Etna, Italy	11,053
St. Elias, Alaska-Yukon	18,008	Lassen Peak, California	10,457
Popocatepetl, Mexico	17,887	Kosciusko, Australia	7,316
Dykh-Tau, U.S.S.R.	17,070	Mitchell, North Carolina	6,684

LONGEST RIVERS OF THE WORLD

	LENGTH IN MILES		LENGTH IN MILES
Nile, Africa	4,145	Japura, S.A.	1,500
Amazon, S.A.	3,915	Arkansas, U.S.A.	1,450
Mississippi-Missouri, U.S.A.	3,710	Colorado, U.S.A.-Mexico	1,450
Yangtze, China	3,434	Negro, S.A.	1,400
Ob-Irtysh, U.S.S.R.	3,362	Dnieper, U.S.S.R.	1,368
Yenisey-Angara, U.S.S.R.	3,100	Irrawaddy, Burma	1,325
Hwang (Yellow), China	2,903	Orange, Africa	1,350
Amur, Asia	2,744	Ohio-Allegheny, U.S.A.	1,306
Lena, U.S.S.R.	2,734	Kama, U.S.S.R.	1,262
Congo, Africa	2,718	Columbia, U.S.A.-Canada	1,243
Mackenzie-Peace Canada	2,635	Red, U.S.A.	1,222
Mekong, Asia	2,600	Don, U.S.S.R.	1,222
Niger, Africa	2,585	Brazos, U.S.A.	1,210
Parana, S.A.	2,450	Saskatchewan, Canada	1,205
Murray-Darling, Australia	2,310	Peace-Finlay, Canada	1,195
Volga, U.S.S.R.	2,194	Tigris, Asia	1,181
Madeira, S.A.	2,013	Darling, Australia	1,160
Purus, S.A.	1,995	Angara, U.S.S.R.	1,135
Yukon, Alaska-Canada	1,979	Sungari, Asia	1,130
St. Lawrence, Canada-U.S.A.	1,900	Pechora, U.S.S.R.	1,124
Rio Grande, U.S.A.-Mexico	1,885	Snake, U.S.A.	1,038
Syr-Dar'ya, U.S.S.R.	1,859	Churchill, Canada	1,000
Sao Francisco, Brazil	1,811	Pilcomayo, S.A.	1,000
Indus, Asia	1,800	Uruguay, S.A.	1,000
Danube, Europe	1,775	Magdalena, Colombia	1,000
Salween, Asia	1,770	Platte-N. Platte, U.S.A.	990
Brahmaputra, Asia	1,700	Oka, U.S.S.R.	918
Euphrates, Asia	1,700	Canadian, U.S.A.	906
Tocantins, Brazil	1,677	Tennessee, U.S.A.	900
Si, China	1,650	Colorado, Texas, U.S.A.	894
Amu-Dar'ya, Asia	1,616	Dniester, U.S.S.R.	876
Zambezi, Africa	1,600	South Saskatchewan, Canada	865
Nelson, Canada	1,600	Fraser, Canada	850
Orinoco, S.A.	1,600	Rhine, Europe	820
Paraguay, S.A.	1,584	Northern Dvina, U.S.S.R.	809
Kolyma, U.S.S.R.	1,562	Tisza, Europe	800
Ganges, Asia	1,550	North Canadian, U.S.A.	784
Ural, U.S.S.R.	1,509	Athabasca, Canada	765

Between Principal Cities in the United States

FROM/TO	Albuquerque, N. Mex.	Atlanta, Ga.	Baltimore, Md.	Boise, Idaho	Boston, Mass.	Brownsville, Tex.	Buffalo, N. Y.	Chicago, Ill.	Cincinnati, Ohio	Cleveland, Ohio	Denver, Colo.	Des Moines, Iowa	Detroit, Mich.	El Paso, Tex.	Fargo, N. Dak.	Fort Worth, Tex.	Galveston, Tex.	Hastings, Nebr.	Hot Springs, Ark.	Houghton, Mich.
Albuquerque, N. Mex.	....	1273	1670	774	1967	838	1577	1126	1248	1417	332	833	1360	228	968	561	803	588	773	1252
Atlanta, Ga.	1273	....	575	1830	933	960	695	583	368	550	1208	738	595	1293	1112	750	688	901	498	947
Baltimore, Md.	1670	575	....	2055	358	1525	273	603	423	305	1505	913	398	1750	1143	1239	1245	1154	964	808
Boise, Idaho	774	1830	2055	....	2266	1610	1872	1453	1663	1754	637	1155	1671	969	975	1263	1538	934	1384	1367
Boston, Mass.	1967	933	358	2266	....	1881	398	849	737	550	1766	1159	613	2067	1304	1574	1598	1415	1302	922
Brownsville, Tex.	838	960	1525	1610	1881	....	1575	1234	1184	1402	1047	1102	1398	682	1445	471	287	1013	650	1543
Buffalo, N. Y.	1577	695	273	1872	398	1575	....	454	392	175	1368	762	218	1690	923	1221	1289	1019	956	560
Chicago, Ill.	1126	583	603	1453	849	1234	454	....	249	307	918	310	236	1249	571	820	954	566	585	367
Cincinnati, Ohio	1248	368	423	1663	737	1184	392	249	....	218	1090	509	234	1333	818	839	897	742	569	589
Cleveland, Ohio	1417	550	305	1754	550	1402	175	307	218	....	1223	617	94	1521	838	1046	1116	871	787	518
Denver, Colo.	332	1208	1505	637	1766	1047	1368	918	1090	1223	....	607	1153	554	642	643	925	353	749	970
Des Moines, Iowa	833	738	913	1155	1159	1102	762	310	509	617	607	....	545	980	397	640	851	256	488	458
Detroit, Mich.	1360	595	398	1671	613	1398	218	236	234	94	1153	545	....	1475	745	1018	1111	800	761	427
El Paso, Tex.	228	1293	1750	969	2067	682	1690	1249	1333	1521	554	980	1475	....	1161	543	723	757	802	1422
Fargo, N. Dak.	968	1112	1143	975	1304	1445	923	571	818	838	642	397	745	1161	....	973	1218	440	875	393
Fort Worth, Tex.	561	750	1239	1263	1574	471	1221	820	839	1046	643	640	1018	543	973	....	283	544	273	1093
Galveston, Tex.	803	688	1245	1538	1598	287	1289	954	897	1116	925	851	1111	723	1218	283	....	808	375	1277
Hastings, Nebr.	588	901	1154	934	1415	1013	1019	566	742	871	353	256	800	757	440	544	808	....	513	666
Hot Springs, Ark.	773	498	964	1384	1302	650	956	585	569	787	749	488	761	802	875	273	375	513	....	901
Houghton, Mich.	1252	947	808	1367	922	1543	560	367	589	518	970	458	427	1422	393	1093	1277	666	901	....
Jacksonville, Fla.	1492	286	682	2098	1015	1025	880	861	628	768	1468	1024	832	1481	1400	943	799	1178	728	1216
Kansas City, Mo.	717	675	962	1158	1250	923	862	413	541	700	555	180	643	836	548	460	677	226	326	633
Los Angeles, Calif.	663	1935	2313	663	2590	1370	2195	1741	1892	2044	828	1433	1976	702	1426	1212	1423	1177	1437	1787
Louisville, Ky.	1174	317	498	1623	823	1093	483	268	92	309	1035	477	315	1253	818	751	807	693	480	636
Memphis, Tenn.	938	335	792	1506	1133	777	802	481	410	627	878	485	621	978	882	448	492	591	176	830
Miami, Fla.	1710	610	958	2368	1258	1100	1184	1190	957	1088	1732	1338	1156	1662	1721	1150	941	1468	983	1545
Minneapolis, Minn.	980	905	948	1140	1125	1335	733	356	603	632	699	235	542	1156	219	870	1087	399	722	272
Missoula, Mont.	895	1790	1947	252	2124	1706	1740	1348	1578	1640	670	1074	1552	1115	819	1312	1595	891	1385	1208
Nashville, Tenn.	1117	218	597	1631	941	952	626	394	239	456	1018	523	468	1169	900	643	666	697	370	760
New Orleans, La.	1030	427	1001	1713	1359	536	1087	831	708	922	1079	825	938	986	1221	470	288	870	358	1187
New York, N. Y.	1810	747	170	2153	188	1695	291	711	568	404	1628	1023	483	1902	1213	1398	1415	1275	1125	849
Norfolk, Va.	1696	507	167	2137	467	1465	435	696	474	429	1562	983	522	1755	1258	1226	1195	1216	955	946
Oklahoma, Okla.	518	753	1173	1138	1490	659	1117	689	755	946	503	469	905	578	786	188	456	357	260	926
Omaha, Nebr.	718	815	1026	1044	1280	1061	883	432	620	738	485	122	666	875	390	590	828	135	490	547
Philadelphia, Pa.	1748	663	90	2113	268	1614	278	664	501	343	575	972	444	1834	1186	1324	1335	1222	1051	827
Phoenix, Ariz.	330	1592	2002	733	2295	1023	1904	1451	1578	1745	585	1154	1685	347	1225	858	1065	901	1094	1550
Pittsburgh, Pa.	1498	520	194	1863	478	1424	178	411	258	115	1320	718	208	1592	952	1097	1140	967	825	530
Portland, Me.	2015	1022	446	2282	100	1961	438	892	802	603	1803	1197	657	2126	1313	1642	1678	1454	1371	924
Portland, Oreg.	1107	2172	2367	349	2553	1944	2167	1765	1987	2063	985	1479	1975	1286	1248	1612	1885	1271	1733	1638
Richmond, Va.	1628	470	128	2060	471	1428	375	618	399	353	1488	905	445	1695	1180	1170	1154	1142	897	870
St. Louis, Mo.	938	467	731	1389	1036	975	662	259	308	490	793	270	452	1033	658	568	697	455	325	591
Salt Lake City, Utah	483	1580	1858	292	2099	1317	1701	1260	1450	1567	372	952	1490	689	865	977	1249	708	1116	1242
San Francisco, Calif.	893	2133	2451	516	2696	1675	2298	1855	2037	2163	946	1547	2087	993	1447	1454	1693	1297	1648	1833
Schenectady, N. Y.	1823	840	278	2120	150	1770	249	702	605	408	1618	1012	467	1930	1157	1445	1487	1267	1175	776
Seattle, Wash.	1178	2180	2341	405	2508	2015	2130	1743	1974	2035	1020	1470	1945	1373	1206	1658	1938	1288	1759	1588
Shreveport, La.	764	548	1064	1433	1410	510	1080	725	688	904	799	624	891	752	1002	209	233	615	142	1043
Spokane, Wash.	1028	1960	2110	290	2279	1852	1900	1514	1746	1804	827	1243	1715	1238	976	1470	1753	1061	1552	1360
Springfield, Mass.	1889	863	282	2196	79	1805	325	774	659	473	1692	1085	540	1990	1240	1495	1524	1340	1224	860
Vermillion, S. Dak.	742	917	1083	973	1314	1161	916	479	694	785	468	187	705	920	284	689	938	167	605	510
Washington, D. C.	1648	542	33	2045	392	1493	290	594	403	303	1490	895	397	1726	1141	1210	1214	1139	936	813

FROM/TO	Jacksonville, Fla.	Kansas City, Mo.	Los Angeles, Calif.	Louisville, Ky.	Memphis, Tenn.	Miami, Fla.	Minneapolis, Minn.	Missoula, Mont.	Nashville, Tenn.	New Orleans, La.	New York, N. Y.	Norfolk, Va.	Oklahoma, Okla.	Omaha, Nebr.	Philadelphia, Pa.	Phoenix, Ariz.	Pittsburgh, Pa.
Albuquerque, N. Mex.	1492	717	663	1174	938	1710	980	895	1117	1030	1810	1696	518	718	1748	330	1498
Atlanta, Ga.	286	675	1935	317	335	610	905	1790	218	427	747	507	753	815	663	1592	520
Baltimore, Md.	682	962	2313	498	792	958	948	1947	597	1001	170	167	1173	1026	90	2002	194
Boise, Idaho	2098	1158	663	1623	1506	2368	1140	252	1631	1713	2153	2137	1138	1044	2113	733	1863
Boston, Mass.	1015	1250	2590	823	1133	1258	1125	2124	941	1359	188	467	1490	1280	268	2295	478
Brownsville, Tex.	1025	923	1370	1093	777	1100	1335	1706	952	536	1695	1465	659	1061	1614	1023	1424
Buffalo, N. Y.	880	862	2195	483	802	1184	733	1740	626	1087	291	435	1117	883	278	1904	178
Chicago, Ill.	861	413	1741	268	481	1190	356	1348	394	831	711	696	689	432	664	1451	411
Cincinnati, Ohio	628	541	1892	92	410	957	603	1578	239	708	568	474	755	620	501	1578	258
Cleveland, Ohio	768	700	2044	309	627	1088	632	1640	456	922	404	429	946	738	343	1745	115
Denver, Colo.	1468	555	828	1035	878	1732	699	670	1018	1079	1628	1562	503	485	1575	585	1320
Des Moines, Iowa	1024	180	1433	477	485	1338	235	1074	523	825	1023	983	469	122	972	1154	718
Detroit, Mich.	832	643	1976	315	621	1156	542	1552	468	938	483	522	905	666	444	1685	208
El Paso, Tex.	1481	836	702	1253	978	1662	1156	1115	1169	986	1902	1755	578	875	1834	347	1592
Fargo, N. Dak.	1400	548	1426	818	882	1721	219	819	900	1221	1213	1258	786	390	1186	1225	952
Fort Worth, Tex.	943	460	1212	751	448	1150	870	1312	643	470	1398	1226	188	590	1324	858	1097
Galveston, Tex.	799	677	1423	807	492	941	1087	1595	666	288	1415	1195	456	828	1336	1065	1140
Hastings, Nebr.	1178	226	1177	693	591	1468	399	891	697	870	1275	1216	357	135	1222	901	967
Hot Springs, Ark.	728	326	1437	480	176	983	722	1385	370	358	1125	955	260	490	1051	1094	825
Houghton, Mich.	1216	633	1787	636	830	1545	272	1208	760	1187	849	946	926	547	827	1550	630
Jacksonville, Fla.	....	952	2153	595	591	328	1192	2070	502	511	838	548	988	1098	758	1800	703
Kansas City, Mo.	952	....	1352	480	370	1247	413	1117	472	678	1097	1009	293	165	1037	1045	784
Los Angeles, Calif.	2153	1352	....	1825	1602	2355	1522	910	1777	1675	2446	2352	1182	1312	2388	357	2135
Louisville, Ky.	595	480	1825	....	319	923	605	1550	153	623	650	528	675	579	580	1512	345
Memphis, Tenn.	591	370	1602	319	....	878	700	1483	195	358	953	778	422	529	878	1264	660
Miami, Fla.	328	1247	2355	923	878	....	1516	2359	821	681	1095	802	1233	1402	1023	1998	1014
Minneapolis, Minn.	1192	413	1522	605	700	1516	....	1010	695	1050	1019	1047	692	291	985	1279	745
Missoula, Mont.	2070	1117	910	1550	1483	2359	1010	....	1582	1733	2030	2045	1162	978	1997	932	1754
Nashville, Tenn.	502	472	1777	153	195	821	695	1582	....	470	758	586	602	604	683	1445	472
New Orleans, La.	511	678	1675	623	358	681	1050	1733	470	....	1173	932	575	845	1090	1318	923
New York, N. Y.	838	1097	2446	650	953	1095	1019	2030	758	1173	....	293	1324	1144	83	2142	313
Norfolk, Va.	548	1009	2352	528	778	802	1047	2045	586	932	293	....	1186	1095	220	2027	316
Oklahoma, Okla.	988	293	1182	675	422	1233	692	1162	602	575	1324	1186	....	405	1256	843	1013
Omaha, Nebr.	1098	165	1312	579	529	1402	291	978	604	845	1144	1095	405	....	1094	1032	837
Philadelphia, Pa.	758	1037	2388	580	878	1023	985	1997	683	1090	83	220	1256	1094	....	2079	254
Phoenix, Ariz.	1800	1045	357	1512	1264	1998	1279	932	1445	1318	2142	2027	843	1032	2079	....	1829
Pittsburgh, Pa.	703	784	2135	345	660	1014	745	1754	472	923	313	316	1013	837	254	1829	....
Portland, Me.	1113	1300	2631	892	1205	1357	1145	2133	1015	1445	277	565	1550	1318	360	2345	545
Portland, Oreg.	2442	1397	825	1953	1852	2716	1435	430	1970	2063	2455	2458	1488	1373	2419	1007	2174
Richmond, Va.	953	937	2283	457	722	831	968	1967	526	899	287	79	1122	1020	205	1960	242
St. Louis, Mo.	755	238	1585	242	242	1067	464	1331	253	599	873	771	456	352	808	1270	561
Salt Lake City, Utah	1840	922	577	1400	1250	2098	988	435	1390	1433	1972	1925	862	833	1923	504	1670
San Francisco, Calif.	2375	1500	345	1983	1800	2603	1585	762	1958	1923	2568	2510	1386	1425	2518	652	2264
Schenectady, N. Y.	960	1107	2445	695	1010	1229	975	1978	820	1259	142	426	1354	1133	205	2152	350
Seattle, Wash.	2450	1505	956	1945	1867	2740	1403	395	1973	2098	2419	2440	1523	1372	2388	1112	2145
Shreveport, La.	733	326	1420	598	279	950	859	1457	470	280	1230	1037	297	617	1153	1067	939
Spokane, Wash.	2239	1286	939	1720	1652	2528	1173	170	1752	1898	2190	2211	1324	1149	2159	1020	1918
Springfield, Mass.	957	1173	2515	745	1055	1210	1056	2060	863	1287	120	411	1412	1205	201	2220	400
Vermillion, S. Dak.	1203	280	1291	663	642	1510	238	887	704	960	1189	1166	502	115	1143	1043	891
Washington, D. C.	647	943	2295	473	763	927	936	1940	567	968	204	145	1150	1012	122	1980	188

Between Principal Cities of Europe

FROM/TO	Amsterdam	Athens	Baku	Barcelona	Belgrade	Berlin	Brussels	Bucharest	Budapest	Cologne	Copenhagen	Istanbul	Dresden	Dublin	Frankfurt	Hamburg	Leningrad	Lisbon	London	Lyon
Amsterdam	....	1340	2218	770	875	365	105	1100	710	128	381	1360	385	468	228	232	1090	1140	220	458
Athens	1340	....	1395	1160	500	1112	1292	460	698	1200	1320	350	1022	1765	1113	1250	1535	1770	1476	1100
Baku	2218	1395	....	2427	1487	1867	2240	1220	1562	2127	1980	1070	1837	2490	2055	2020	1570	3050	2435	2238
Barcelona	770	1160	2427	....	998	925	658	1210	924	692	1085	1380	860	919	665	910	1740	610	707	327
Belgrade	875	500	1487	998	....	618	850	295	205	750	840	502	530	1327	652	760	1165	1555	1040	752
Berlin	365	1112	1867	925	618	....	401	798	425	300	225	1068	95	815	268	165	815	1410	575	601
Brussels	105	1292	2240	658	850	401	....	1110	700	110	475	1345	407	480	198	301	1175	998	202	352
Bucharest	1100	460	1220	1210	295	798	1110	....	295	982	970	272	725	1560	890	950	1080	1842	1285	1025
Budapest	710	698	1562	924	205	425	700	295	....	590	629	650	345	1176	504	572	965	1515	900	680
Cologne	128	1200	2127	692	750	300	110	982	590	....	400	1240	292	585	93	228	1090	1126	308	370
Copenhagen	381	1320	1960	1085	840	225	475	970	629	400	....	1240	315	768	412	180	708	1520	590	760
Istanbul	1360	350	1070	1380	502	1068	1345	272	650	1240	1240	....	995	1830	1150	1222	1292	2005	1540	1238
Dresden	385	1022	1837	860	530	95	407	725	345	292	315	995	....	852	236	238	885	1380	592	540
Dublin	468	1765	2490	919	1327	815	480	1560	1176	585	768	1830	852	....	671	668	1440	1015	300	720
Frankfurt	228	1113	2055	665	652	268	198	890	504	93	412	1150	236	671	....	250	1075	1160	392	350
Hamburg	232	1250	2020	910	760	165	301	950	572	228	180	1222	238	668	250	....	880	1301	448	580
Leningrad	1090	1535	1570	1740	1165	815	1175	1080	965	1090	708	1292	885	1440	1075	880	....	2235	1300	1420
Lisbon	1140	1770	3050	610	1555	1410	998	1842	1515	1126	1520	2005	1380	1015	1160	1301	2235	....	975	850
London	220	1476	2435	707	1040	575	202	1285	900	308	590	1540	592	300	392	448	1300	975	....	455
Lyon	458	1100	2238	327	752	601	352	1025	680	370	760	1238	540	720	350	580	1420	850	455	....
Madrid	912	1463	2742	316	1235	1149	807	1518	1214	875	1272	1690	1100	902	888	1098	1980	313	777	557
Marseille	627	1025	2238	211	750	730	521	1020	718	528	906	1205	655	875	492	730	1540	810	620	170
Milan	517	900	2028	450	540	570	435	819	476	390	720	1030	435	880	323	570	1315	1350	595	210
Moscow	1325	1388	1175	1852	1160	995	1392	920	965	1285	970	1180	1200	1728	1240	1100	391	430	1540	1560
Munich	415	925	1912	648	475	310	372	725	350	282	520	975	227	855	193	378	1100	1208	526	352
Oslo	568	1610	2118	1330	1112	520	672	1245	920	635	303	1505	620	786	675	445	670	1690	720	1005
Paris	257	1300	2335	518	890	540	170	1152	770	250	634	1390	523	480	295	459	1335	890	210	248
Riga	820	1310	1590	1440	855	520	900	870	685	805	453	1115	585	1210	780	600	300	1940	1035	1122
Rome	808	650	1900	530	440	730	730	700	500	675	948	840	630	1175	698	810	1440	1150	890	462
Sofia	1073	335	1360	1072	231	810	945	194	395	945	1010	315	730	1525	860	954	1218	1685	1235	928
Stockholm	695	1495	1862	1410	1005	503	793	1080	820	722	330	1340	598	1010	730	502	435	1848	885	1080
Toulouse	625	1215	2425	156	930	815	515	1210	883	875	962	1400	762	761	560	780	1635	640	550	228
Warsaw	673	990	1555	1150	510	320	720	580	342	602	415	852	325	1130	550	462	640	1700	890	850
Vienna	580	795	1700	830	300	322	568	520	128	460	538	790	235	1040	370	460	975	1415	762	562
Zürich	375	1000	2050	513	590	410	312	855	498	259	595	1090	342	768	193	432	1225	1058	480	206

FROM/TO	Madrid	Marseille	Milan	Moscow	Munich	Oslo	Paris	Riga	Rome	Sofia	Stockholm	Toulouse	Warsaw	Vienna	Zürich
Amsterdam	912	627	517	1325	415	568	257	820	808	1073	695	625	673	580	375
Athens	1463	1025	900	1388	925	1610	1300	1310	650	335	1495	1215	990	795	1000
Baku	2742	2238	2028	1175	1912	2118	2335	1590	1900	1360	1862	2425	1555	1700	2050
Barcelona	316	211	450	1852	648	1330	518	1440	530	1072	1410	156	1150	830	513
Belgrade	1235	750	540	1160	475	1112	890	855	440	231	1005	930	510	300	590
Berlin	1149	730	570	995	310	520	540	520	730	810	503	815	320	322	410
Brussels	807	521	435	1392	372	672	170	900	730	945	793	515	720	568	312
Bucharest	1518	1020	819	920	725	1245	1152	870	700	194	1080	1210	580	520	855
Budapest	1214	718	476	965	350	920	770	685	500	395	820	883	342	128	498
Cologne	875	528	390	1285	282	635	250	805	675	945	722	875	602	460	259
Copenhagen	1272	906	720	970	520	303	634	453	948	1010	330	962	415	538	595
Istanbul	1690	1205	1030	1180	975	1505	1390	1115	840	315	1340	1400	852	790	1090
Dresden	1100	655	435	1200	227	620	523	585	630	730	598	762	325	235	342
Dublin	902	875	880	1728	855	786	480	1210	1175	1525	1010	761	1130	1040	768
Frankfurt	888	492	323	1240	193	675	295	780	698	860	730	560	550	370	193
Hamburg	1098	730	570	1100	378	445	459	600	810	954	502	780	462	460	432
Leningrad	1980	1540	1315	391	1100	670	1335	300	1440	1218	435	1635	640	975	1225
Lisbon	313	810	1350	430	1208	1690	890	1940	1150	1685	1848	640	1700	1415	1058
London	777	620	595	1540	526	720	210	1035	890	1235	885	550	890	762	480
Lyon	577	170	210	1560	352	1005	248	1122	462	928	1080	228	850	562	206
Madrid	....	394	728	2120	910	1474	645	1670	840	1385	1598	344	1410	1110	765
Marseille	394	....	238	1642	445	1165	410	1238	372	895	1225	196	950	620	318
Milan	728	238	....	1408	215	1000	400	1010	295	715	1020	400	705	385	137
Moscow	2120	1642	1408	....	1220	1030	1538	520	1462	1100	770	1770	710	1028	1350
Munich	910	445	215	1220	....	810	425	800	430	672	811	570	500	222	158
Oslo	1474	1165	1000	1030	810	....	830	531	1242	1295	267	1140	653	835	869
Paris	645	410	400	1538	425	830	....	1050	690	1080	950	431	845	770	295
Riga	1670	1238	1010	520	800	531	1050	....	1155	985	276	1335	350	685	930
Rome	840	372	295	1462	430	1242	690	1155	....	545	1220	569	810	470	421
Sofia	1385	895	715	1100	672	1295	1080	985	545	....	1170	1080	662	500	780
Stockholm	1598	1225	1020	770	811	267	950	276	1220	1170	....	1281	500	770	908
Toulouse	344	196	400	1770	570	1140	431	1335	569	1080	1281	....	1062	725	425
Warsaw	1410	950	705	710	500	653	845	350	810	662	500	1062	....	345	640
Vienna	1110	620	385	1028	222	835	770	685	470	500	770	725	345	....	365
Zürich	765	318	137	1350	158	869	295	930	421	780	908	425	640	365	....

AIRLINE DISTANCES

All Distances in Statute Miles

Richmond, Va.	St. Louis, Mo.	Salt Lake City, Utah	San Francisco, Calif.	Schenectady, N. Y.	Seattle, Wash.	Shreveport, La.	Spokane, Wash.	Springfield, Mass.	Vermillion, S. Dak.	Washington, D.
1628	938	483	893	1823	1178	764	1028	1889	742	1648
470	467	1580	2133	840	2180	548	1960	863	917	542
128	731	1858	2451	278	2341	1064	2110	282	1083	33
2060	1389	292	516	2120	405	1433	290	2196	973	2045
471	1036	2099	2696	150	2508	1410	2279	79	1314	392
1428	975	1317	1675	1770	2015	510	1852	1805	1161	1493
375	662	1701	2298	249	2130	1080	1900	325	916	290
618	259	1260	1855	702	1743	725	1514	774	479	594
399	308	1450	2037	605	1974	688	1746	659	694	403
353	490	1567	2163	408	2035	904	1804	478	785	303
1488	793	372	946	1618	1020	799	827	1692	468	1490
905	270	952	1547	1012	1470	624	1243	1085	187	895
445	452	1490	2087	467	1945	891	1715	540	705	397
1695	1033	689	993	1930	1373	752	1238	1990	920	1726
1180	658	865	1447	1157	1206	1002	976	1240	284	1141
1170	568	977	1454	1445	1658	209	1470	1495	689	1210
1154	697	1249	1693	1487	1938	233	1753	1524	938	1214
1142	455	708	1297	1267	1288	615	1061	1340	167	1139
897	325	1116	1648	1175	1759	142	1552	1224	605	936
870	591	1242	1833	776	1588	1043	1360	860	510	813
953	755	1840	2375	960	2450	733	2239	957	1203	647
937	238	922	1500	1107	1505	326	1286	1173	280	943
2283	1585	577	345	2445	956	1420	939	2515	1291	2295
457	242	1400	1983	695	1945	598	1720	745	663	473
722	242	1250	1800	1010	1867	279	1652	1055	642	763
831	1067	2098	2603	1229	2740	950	2528	1210	1510	927
968	464	988	1585	975	1403	859	1173	1056	238	936
1967	1331	435	762	1978	395	1457	170	2060	887	1940
526	253	1390	1958	820	1973	470	1752	863	704	567
899	599	1433	1923	1259	2098	280	1898	1287	960	968
287	873	1972	2568	142	2419	1230	2190	120	1189	204
79	771	1925	2510	426	2440	1037	2211	411	1166	145
1122	456	862	1386	1354	1523	297	1324	1412	502	1150
1020	352	833	1425	1133	1372	617	1149	1205	115	1012
205	808	1923	2518	205	2388	1153	2159	201	1143	122
1960	1270	504	652	2152	1112	1067	1020	2220	1043	1980
242	561	1670	2264	350	2145	939	1918	400	891	188
565	1094	2127	2725	197	2513	1484	2285	159	1345	480
2381	1723	636	536	2405	143	1783	295	2488	1293	2360
....	699	1850	2436	406	2362	985	2133	407	1089	96
699	....	1158	1738	898	1722	466	1500	958	450	710
1850	1158	....	592	1950	697	1155	548	2027	785	1845
2436	1738	592	....	2548	680	1655	730	2625	1383	2437
406	898	1950	2548	....	2363	1290	2139	86	1165	313
2362	1722	697	680	2363	....	1820	229	2445	1282	2335
985	466	1155	1655	1290	1820	....	1621	1333	726	1035
2133	1500	548	730	2139	229	1621	....	2216	1055	2105
407	958	2027	2625	86	2445	1333	2216	....	1242	321
1089	450	785	1383	1165	1282	726	1055	1242	....	1073
96	710	1845	2437	313	2335	1035	2105	321	1073	....

Between Representative Cities of the United States and Latin America

NEW YORK TO	Miles	SAN FRANCISCO TO	Miles	SEATTLE TO	Miles	WASHINGTON TO	Miles
Belém	3,281	Belém	5,430	Belém	5,550	Belém	3,270
Buenos Aires	5,295	Buenos Aires	6,487	Buenos Aires	6,956	Buenos Aires	5,205
Bogotá	2,474	Bogotá	3,863	Bogotá	4,166	Bogotá	2,344
Caracas	2,100	Caracas	3,900	Caracas	4,100	Caracas	2,040
Guatemala City	2,060	Guatemala City	2,525	Guatemala City	2,930	Guatemala City	1,835
Havana	1,302	Havana	2,600	Havana	2,805	Havana	1,110
La Paz	3,905	La Paz	5,080	La Paz	5,110	La Paz	3,780
Panamá	2,211	Panamá	3,349	Panamá	3,680	Panamá	2,020
Managua	2,100	Managua	2,860	Managua	3,240	Managua	1,920
Rio de Janeiro	4,810	Rio de Janeiro	6,655	Rio de Janeiro	6,945	Rio de Janeiro	4,710
San José	2,200	San José	3,070	San José	3,430	San José	2,030
Santiago	5,134	Santiago	5,960	Santiago	6,466	Santiago	4,965
Tampico	1,880	Tampico	1,790	Tampico	2,200	Tampico	1,665

CHICAGO TO	Miles	DENVER TO	Miles	LOS ANGELES TO	Miles	NEW ORLEANS TO	Miles
Belém	3,820	Bélem	4,580	Belém	5,110	Belém	3,470
Buenos Aires	5,598	Buenos Aires	5,935	Buenos Aires	6,148	Buenos Aires	4,902
Bogotá	2,691	Bogotá	3,100	Bogotá	3,515	Bogotá	1,996
Caracas	2,480	Caracas	3,105	Caracas	3,610	Caracas	1,990
Guatemala City	1,870	Guatemala City	1,935	Guatemala City	2,190	Guatemala City	1,050
Havana	1,315	Havana	1,760	Havana	2,320	Havana	672
La Paz	4,130	La Paz	4,445	La Paz	4,805	La Paz	3,480
Panamá	2,320	Panamá	2,620	Panamá	3,025	Panamá	1,600
Managua	2,060	Managua	2,230	Managua	2,540	Managua	1,250
Rio de Janeiro	5,320	Rio de Janeiro	5,900	Rio de Janeiro	6,330	Rio de Janeiro	4,798
San José	2,100	San José	2,420	San José	2,725	San José	1,425
Santiago	5,320	Santiago	5,495	Santiago	5,595	Santiago	4,553
Tampico	1,460	Tampico	1,240	Tampico	1,470	Tampico	720

Between Principal Cities of the World

FROM/TO	Azores	Bagdad	Berlin	Bombay	Buenos Aires	Callao	Cairo	Cape Town	Chicago	Istanbul	Guam	Honolulu	Juneau	London	Los Angeles
Azores	....	3906	2148	5930	5385	4825	3325	5670	3305	2880	8985	7421	4715	1562	5034
Bagdad	3906	....	2040	2022	8215	8618	785	4923	6490	1085	6380	8445	6180	2568	7695
Berlin	2148	2040	....	3947	7411	6937	1823	5949	4458	1068	7158	7384	4638	575	5849
Bombay	5930	2022	3947	....	9380	10530	2698	5133	8144	3043	4831	8172	6992	4526	8810
Buenos Aires	5385	8215	7411	9380	....	1982	7428	4332	5598	7638	10516	7653	7964	6919	6148
Callao	4825	8618	6937	10530	1982	....	7870	6195	3765	7666	9760	5993	5806	6376	4155
Cairo	3325	785	1823	2698	7428	7870	....	4476	6231	780	7175	8925	6352	2218	7675
Cape Town	5670	4923	5949	5133	4332	6195	4476	....	8551	5210	8918	11655	10382	5975	10165
Chicago	3305	6490	4458	8144	5598	3765	6231	8551	....	5530	7510	4315	2310	4015	1741
Istanbul	2880	1085	1068	3043	7638	7666	780	5210	5530	....	7015	8200	5665	1540	6895
Guam	8985	6380	7158	4831	10516	9760	7175	8918	7510	7015	....	3896	5225	7605	6255
Honolulu	7421	8445	7384	8172	7653	5993	8925	11655	4315	8200	3896	....	2825	7320	2620
Juneau	4715	6180	4638	6992	7964	5806	6352	10382	2310	5665	5225	2825	....	4496	1835
London	1562	2568	575	4526	6919	6376	2218	5975	4015	1540	7605	7320	4496	....	5496
Los Angeles	5034	7695	5849	8810	6148	4155	7675	10165	1741	6895	6255	2620	1835	5496	....
Melbourne	12190	8150	9992	6140	7336	8196	8720	6510	9837	9189	3497	5581	8162	10590	8098
Mexico City	4584	8155	6119	9818	4609	2619	7807	8620	1690	7160	7690	3846	3210	5605	1445
Montréal	2548	5814	3776	7582	5619	3954	5502	7975	750	4825	7840	4992	2647	3370	2468
New Orleans	3718	7212	5182	8952	4902	2990	6862	8390	827	6220	7895	4305	2860	4656	1695
New York	2604	6066	4025	7875	5295	3633	5701	7845	727	5060	8115	5051	2874	3500	2466
Panamá	3918	7807	5902	9832	3319	1450	7230	7090	2320	6797	9220	5347	4456	5310	3025
Paris	1617	2385	540	4391	6891	6455	2020	5762	4219	1390	7675	7525	4700	210	5711
Rio de Janeiro	4312	7012	6246	8438	1230	2400	6242	3850	5320	6420	11710	8400	7611	5747	6330
San Francisco	5114	7521	5744	8523	6487	4500	7554	10340	1875	6770	5952	2407	1530	5440	345
Santiago	5718	8876	7842	10127	731	1548	8100	5080	5325	8230	9946	6935	7320	7275	5595
Seattle	4720	6848	5121	7830	6956	4964	6915	10305	1753	6124	5785	2707	870	4850	961
Shanghai	7324	4468	5323	3219	12295	10760	5290	8179	7155	5084	1945	5009	4968	5841	6598
Singapore	8338	4443	6226	2425	9940	11700	5152	6025	9475	5440	2990	6874	7375	6818	8955
Tokyo	7370	5242	5623	4247	11601	9740	6005	9234	6410	5649	1596	3940	4117	6050	5600
Wellington	11475	9782	11384	7752	6341	6696	10360	7149	8465	10790	4206	4676	7501	11790	6806

FROM/TO	Melbourne	Mexico City	Montréal	New Orleans	New York	Panamá	Paris	Rio de Janeiro	San Francisco	Santiago	Seattle	Shanghai	Singapore	Tokyo	Wellington
Azores	12190	4584	2548	3718	2604	3918	1617	4312	5114	5718	4720	7324	8338	7370	11475
Bagdad	8150	8155	5814	7212	6066	7807	2385	7012	7521	8876	6848	4468	4443	5242	9782
Berlin	9992	6119	3776	5182	4026	5902	540	6246	5744	7842	5121	5323	6226	5623	11384
Bombay	6140	9818	7582	8952	7875	9832	4391	8438	8523	10127	7830	3219	2425	4247	7752
Buenos Aires	7336	4609	5619	4902	5295	3319	6891	1230	6487	731	6956	12295	9940	11601	6341
Callao	8196	2619	3954	2990	3633	1450	6455	2400	4500	1548	4964	10760	11700	9740	6696
Cairo	8720	7807	5502	6862	5701	7230	2020	6242	7554	8100	6915	5290	5152	6005	10360
Cape Town	6510	8620	7975	8390	7845	7090	5732	3850	10340	5080	10305	8179	6025	9234	7149
Chicago	9837	1690	750	827	727	2320	4219	5320	1875	5325	1753	7155	9475	6410	8465
Istanbul	9189	7160	4825	6220	5060	6797	1390	6420	6770	8230	6124	5084	5440	5649	10790
Guam	3497	7690	7840	7895	8115	9220	7675	11710	5952	9946	5785	1945	2990	1596	4206
Honolulu	5581	3846	4992	4305	5051	5347	7525	8400	2407	6935	2707	5009	6874	3940	4676
Juneau	8162	3210	2647	2860	2874	4456	4700	7611	1530	7320	870	4968	7375	4117	7501
London	10590	5605	3370	4656	3500	5310	210	5747	5440	7275	4850	5841	6818	6050	11790
Los Angeles	8098	1445	2468	1695	2466	3025	5711	6330	345	5595	961	6598	8955	5600	6806
Melbourne	....	8599	10553	9455	10541	9211	10500	8340	7970	7130	8330	4967	3768	5172	1655
Mexico City	8599	....	2247	940	2110	1532	5800	4810	1870	4122	2339	8120	10495	7190	7003
Montréal	10553	2247	....	1390	340	2545	3490	5110	2557	5461	2309	7141	9280	6546	9206
New Orleans	9455	940	1390	....	1161	1600	4846	4798	1960	4553	2137	7830	10255	6993	7950
New York	10541	2110	340	1161	....	2211	3600	4810	2606	5134	2440	7460	9617	6846	9067
Panamá	9211	1532	2545	1600	2211	....	5440	3311	3349	3000	3680	9430	11800	8560	7580
Paris	10500	5800	3490	4846	3600	5440	....	5710	5680	7300	5080	5855	6730	6132	11865
Rio de Janeiro	8340	4810	5110	4798	4810	3311	5710	....	6655	1852	6945	11510	9875	11600	7510
San Francisco	7970	1870	2557	1960	2606	3349	5680	6655	....	5960	692	6245	8440	5250	6800
Santiago	7130	4122	5461	4553	5134	3000	7300	1852	5960	....	6466	11850	10270	10850	5925
Seattle	8330	2339	2309	2137	2440	3680	5080	6945	692	6466	....	5780	8200	4863	7310
Shanghai	4967	8120	7141	7830	7460	9430	5855	11510	6245	11850	5780	....	2395	1095	6080
Singapore	3768	10495	9280	10255	9617	11800	6730	9875	8440	10270	8200	2395	....	3350	5360
Tokyo	5172	7190	6546	6993	6846	8560	6132	11600	5250	10850	4863	1095	3350	....	5730
Wellington	1655	7003	9206	7950	9067	7580	11865	7510	6800	5925	7310	6080	5360	5730	.....

COLOR MAP SECTION

FOREIGN GEOGRAPHICAL TERMS

A. = Arabic Camb. = Cambodian Ch. = Chinese Czech. = Czechoslovakian Dan. = Danish Du. = Dutch Finn. = Finnish Fr. = French Ger. = German Ice. = Icelandic
It. = Italian Jap. = Japanese Mong. = Mongol Nor. = Norwegian Per. = Persian Port. = Portuguese Russ. = Russian Sp. = Spanish Sw. = Swedish Turk. = Turkish

Å	Nor., Sw.	Stream
Aas	Dan., Nor	Hills
Abajo	Sp	Lower
Ada, Adasi	Turk	Island
Altipiano	It	Plateau
Altiplano	Sp	Plateau
Alv, Alf, Elf	Sw	River
Arrecife	Sp	Reef
Asa	Nor., Sw	Hill
Asaga	Turk	Lower
Austral	Sp	Southern
Baai	Du	Bay
Bab	Arabic	Gate or Strait
Bahia	Sp	Bay
Bahr	Arabic	Marsh, Lake, Sea, River
Baia	Port	Bay
Baie	Fr	Bay, Gulf
Baizo	Port	Low
Bakke	Dan	Hill
Bana	Jap	Cape
Bañados	Sp	Marshes
Band	Per	Mt. Range
Barra	Sp	Reef
Bel	Turk	Pass
Belt	Ger	Strait
Ben	Gaelic	Mountain
Bera	Du	Mountain
Berg	Ger., Du	Mountain
Bir	Arabic	Well
Birket	Arabic	Pond
Boca	Sp	Gulf, Inlet
Boğhaz	Turk	Strait
Bolshoi, Bolshaya	Russ	Big
Bolson	Sp	Depression
Bong	Korean	Mountain
Boreal	Sp	Northern
Breen	Nor	Glacier
Bro	Dan., Nor., Sw	Bridge
Bucht	Ger	Bay
Bugt	Dan	Bay
Bukhta	Russ	Bay
Bukit	Malay	Hill, Mountain
Bukt	Nor., Sw	Bay, Gulf
Burnu, Burun	Turk	Cape, Point
By	Dan., Nor., Sw	Town
Cabo	Port., Sp	Cape
Campos	Port	Plains
Canal	Port., Sp	Channel
Cap, Capo	Fr., It	Cape
Cataratas	Sp	Falls
Catena	It	Mt. Range
Catingas	Port	Open Woodlands
Central, Centrale	Fr., It	Middle
Cerrito, Cerro	Sp	Hill
Cerros	Sp	Hills, Mountains
Chai	Turk	River
Chow	Ch	Town of the second rank
Ciénaga	Sp	Swamp
Ciudad	Sp	City
Col	Fr	Pass
Cordillera	Sp	Mt. Range, Mts.
Côte	Fr	Coast
Csatoria	Magyar	Canal
Cuchilla	Sp	Mt. Range
Curiche	Sp	Swamp
Dag, Dagh	Turk	Mountain
Dağlari	Turk	Mt. Range
Dal	Nor., Sw	Valley
Dar	Arabic	Land
Darya	Per	Salt Lake
Dasht	Per	Desert, Plain
Deniz, Denizi	Turk	Sea, Lake
Desierto	Sp	Desert
Détroit	Fr	Strait
Djeziret	Arabic, Turk	Island
Do	Korean	Island
Doi	Thai	Mountain
Eiland	Du	Island
Elv	Dan., Nor	River
Embalse	Sp	Reservoir
Emi	Berber	Mountain
Erg	Arabic	Dune, Desert
Eski	Turk	Old
Est, Este	Fr., Port., Sp	East
Estero	Sp	Estuary, Creek
Estrecho, Estreito	Sp., Port	Strait
Etang	Fr	Pond, Lagoon, Lake
Fedja, Feij	Arabic	Pass
Fiume	It	River
Fjäll	Sw	Mountain
Fjeld, Fjell	Nor	Hills, Mountain
Fjord	Dan., Nor., Sw	Fiord
Fleuve	Fr	River
Fljót	Icelandic	Stream
Fluss	Ger	River
Fokani, Fukani	Arabic	Upper
Fors	Sw	Waterfall
Fos, Foss	Dan., Nor	Waterfall
Fu	Ch	Town of importance
Gamla	Nor	Old
Gamle	Dan	Old
Gata	Jap	Lake
Gawa	Jap	River
Gebel	Arabic	Mountain
Gebergte	Du	Mt. Range
Gebirge	Ger	Mt. Range
Ghubbet	Arabic	Bay
Gobi	Mongol	Desert
Goe	Jap	Pass
Gol	Mongol, Turk	Lake, Stream
Golf	Ger., Du	Gulf
Golfe	Fr	Gulf
Golfo	Sp., It., Port	Gulf
Gölü	Turk	Lake
Gora	Russ	Mountain
Grand, Grande	Fr., Sp	Big
Groot	Du	Big
Gross	Ger	Big
Grosso	It., Port	Big
Guba	Russ	Bay, Gulf
Gunto	Jap	Archipelago
Gunung	Malay	Mountain
Hai	Ch	Sea
Halbinsel	Ger	Peninsula
Hamáda, Hammada	Arabic	Rocky Plateau
Hamn	Sw	Harbor
Hamún	Per	Marsh
Hanto	Jap	Peninsula
Has, Hassi	Arabic	Well
Hav	Dan., Nor., Sw	Sea, Ocean
Havet	Nor	Bay
Havn	Dan., Nor	Harbor
Havre	Fr	Harbor
Higashi, Higasi	Jap	East
Ho	Ch	River
Hochebene	Ger	Plateau
Hoek	Du	Cape
Hoku	Jap	North
Holm	Dan., Nor., Sw	Island
Hory	Czech	Mountains
Hoved	Dan., Nor	Cape, Promontory
Hsien	Ch	Town of the third class
Hu	Ch	Lake
Huk	Dan., Nor., Sw	Point
Hus, Huus	Dan., Nor., Sw	House
Hwang	Ch	Yellow
Ile	Fr	Island
Ilet	Fr	Islet
Ilot	Fr	Islet
Indre	Dan., Nor	Inner
Inferieur, Inferiore	Fr., It	Lower
Inner, Inre	Sw	Inner
Insel	Ger	Island
Irmak	Turk	River
Isla	Sp	Island
Isola	It	Island
Jabal, Jebel	Arabic	Mountains
Järvi	Finn	Lake
Jaure	Sw	Lake
Jezira	Arabic	Island
Jima	Jap	Island
Joki	Finn	River
Kaap	Du	Cape
Kabir, Kebir	Arabic	Big
Kai	Jap	Sea
Kaikyo	Jap	Strait
Kami	Turk	Upper
Kanaal	Du	Canal
Kanal	Russ., Ger	Canal, Channel
Kao	Thai	Mountain
Kap, Kapp	Nor., Sw., Ice	Cape
Kaupunki	Finn	Town
Kawa	Jap	River
Khao	Thai	Mountain
Khrebet	Russ	Mt. Range
Kiang	Ch	River
Kiao	Ch	Point
Kita	Jap	North
Klein	Du., Ger	Small
Klint	Dan	Promontory
Kô	Jap	Lake
Ko	Thai	Island
Koh	Camb., Khmer	Island
Kong	Ch	River
Kop	Du	Peak, Head
Köping	Sw	Market, Borough
Körfez, Körfezi	Turk	Gulf
Kosa	Russ	Spit
Kosui	Jap	Lake
Kraal	Du	Native Village
Kuchuk	Turk	Small
Kuh	Per	Mountain
Kul	Sinkiang Turki	Lake
Kum	Turk	Desert
Kuro	Jap	Black
Laag	Du	Low
Lac	Fr	Lake
Lago	Port., Sp., It	Lake
Lagoa	Port	Lagoon
Laguna	Sp	Lagoon
Lagune	Fr	Lagoon
Lahti	Finn	Bay, Bight
Län	Sw	County
Lilla	Sw	Small
Lille	Dan., Nor	Small
Llanos	Sp	Plains
Loch	Scottish	Lake
Mae Nam	Thai	River
Mali, Malaya	Russ	Small
Man	Korean	Bay
Mar	Sp., Port	Sea
Mare	It	Sea
Medio	Sp	Middle
Meer	Du	Lake
Meer	Ger	Sea
Mer	Fr	Sea
Meridionale	It	Southern
Meseta	Sp	Plateau
Middelst, Midden	Du	Middle
Minami	Jap	Southern
Mir	Per	Mountain
Mis	Russ	Cape
Misaki	Jap	Cape
Mittel	Ger	Middle
Mont	Fr	Mountain
Montagne	Fr	Mountain
Montaña	Sp	Mountains
Monte	Sp., It., Port	Mountain
More	Russ	Sea
Morro	Port., Sp	Mountain, Promontory
Morue	Fr	Hill
Moyen	Fr	Middle
Muong	Siamese	Town
Mys	Russ	Cape
Nada	Jap	Sea
Naka	Jap	Middle
Nam	Burm., Lao	River
Nan	Ch., Jap	South
Nes	Nor	Cape, Point
Nevado	Sp	Snow covered peak
Nieder	Ger	Lower
Nishi, Nisi	Jap	West
Nizhni, Nizhnyaya	Russ	Lower
Njarga	Finn	Peninsula, Promontory
Nong	Thai	Lake
Noord	Du	North
Nor	Mong	Lake
Nord	Fr., Ger	North
Norte	Sp., It., Port	North
Nos	Russ	Cape
Novi, Novaya	Russ	New
Nusa	Malay	Island
Ny, Nya	Nor., Sw	New
O	Jap	Big
Ö	Nor., Sw	Island
Ober	Ger	Upper
Occidental, Occidentale	Sp., It	Western
Odde	Dan	Point
Oeste	Port	West
Ola	Mong	Mountains
Ooster	Du	Eastern
Opper, Over	Du	Upper
Oriental	Sp., Fr	Eastern
Orientale	It	Eastern
Orta	Turk	Middle
Ost	Ger	East
Ostrov	Russ	Island
Ouest	Fr	West
Öy	Nor	Island
Ozero	Russ	Lake
Pampa	Sp	Plain
Pas	Fr	Channel, Strait
Paso	Sp	Pass
Passo	It., Port	Pass
Peh, Pei	Ch	North
Peña	Sp	Rock, Mountain
Penisola	It	Peninsula
Pequeño	Sp	Small
Pereval	Russ	Pass
Peski	Russ	Desert
Petit	Fr	Small
Phu	Lao, Annamese	Mtn.
Pic	Fr	Mountain
Piccolo	It	Small
Pico	Port., Sp	Mountain, Peak
Pik	Russ	Mountain, Peak
Piton	Fr	Mountain, Peak
Planalto	Port	Plateau
Plato	Russ	Plateau
Pointe	Fr	Point
Poluostrov	Russ	Peninsula
Ponta	Port	Point
Presa	Sp	Reservoir
Presqu'île	Fr	Peninsula
Proliv	Russ	Strait
Pulou, Pulo	Malay	Island
Punt	Du	Point
Punta	Sp., It., Port	Point
Qum	Turk	Desert
Rada	Sp	Inlet
Rade	Fr	Bay, Inlet
Ras	Arabic	Cape
Reka	Russ	River
Retto	Jap	Archipelago
Ria	Sp	Estuary
Río	Sp	River
Rivier, Rivière	Du., Fr	River
Rud	Per	River
Saghir	Arabic	Small
Sai	Jap	West
Saki	Jap	Cape
Salar, Salina	Sp	Salt Deposit
Salto	Sp., Port	Falls
San	Ch., Jap., Korean	Hill
Sanmaek	Korean	Mt. Range
Schiereiland	Du	Peninsula
Se	Camb., Khmer	River
See	Ger	Sea, Lake
Selvas	Sp., Port	Woods, Forest
Senp	Sp	Bay, Gulf
Serra	Port	Mts.
Serranía	Sp	Mts.
Seto	Jap	Strait
Settentrionale	It	Northern
Severni, Severnaya	Russ	North
Shan	Ch., Jap	Hill, Mts.
Shang	Ch	Upper
Shatt	Arabic	River
Shima	Jap	Island
Shimo	Jap	Lower
Shin	Jap	Land
Shiro	Jap	White
Shoto	Jap	Islands
Si	Ch	West
Siao	Ch	Small
Sierra	Sp	Mt. Range, Mts.
Sjö	Nor., Sw	Lake, Sea
Sok, Suk, Souk	Arabic, Ar. Fr	Market
Song	Annamese	River
Sopka	Russ	Volcano
Spitze	Ger	Mt. Peak
Sredni, Srednyaya	Russ	Middle
Stad	Dan., Nor., Sw	City
Stari, Staraya	Russ	Old
Step	Russ	Treeless Plain
Straat	Du	Strait
Strasse	Ger	Strait
Stretto	It	Strait
Ström	Dan., Nor., Sw	Sound
Stung	Camb., Khmer	River
Su	Turk	River
Sud, Süd	Sp., Fr., Ger	South
Suido	Jap	Strait, Channel
Sul	Port	South
Sund	Dan., Nor., Sw	Sound
Sungei	Malay	River
Supérieur	Fr	Upper
Superior, Superiore	Sp., It	Upper
Sur	Sp	South
Suyu	Turk	River
Ta	Ch	Big
Tafelland	Du	Plateau
Tagh	Turk	Mt. Range
Take	Jap	Peak, Ridge
Takht	Arabic	Lower
Tal	Ger	Valley
Tandjong, Tanjung	Malay	Cape, Point
Tao	Ch	Island
Tell	Arabic	Hill
Thale	Thai	Sea, Lake
Tind	Nor	Peak
Tö	Jap	East
To	Jap	Island
Toge	Jap	Pass
Trask	Finn	Lake
Tso	Tibetan	Lake
Tugh	Somali	Dry River
Tung	Ch	Eastern
Udjung	Malay	Point
Umi	Jap	Bay
Unter	Ger	Lower
Ura	Jap	Inlet
Val	Fr	Valley
Vatn	Nor	Lake
Vecchio	It	Old
Veld	Du	Plain, Field
Velho	Port	Old
Verkhni	Russ	Upper
Vesi	Finn	Lake
Vieho	Sp	Old
Vik	Nor., Sw	Bay
Vishni, Vishnyaya	Russ	High
Vodokhranilishche	Russ	Reservoir
Volcán	Sp	Volcano
Vostochni, Vostochnaya	Russ	East, Eastern
Wadi	Arabic	Dry River
Wald	Ger	Forest
Wan	Jap	Bay
Westersch	Du	Western
Wüste	Ger	Desert
Yama	Jap	Mountain
Yarim Ada	Turk	Peninsula
Yokara	Turk	Upper
Yug, Yuzhni, Yuzhnaya	Russ	South, Southern
Zaki	Jap	Cape
Zaliv	Russ	Bay, Gulf
Zapadni, Zapadnaya	Russ	Western
Zee	Du	Sea
Zemlya	Russ	Land
Zuid	Du	South

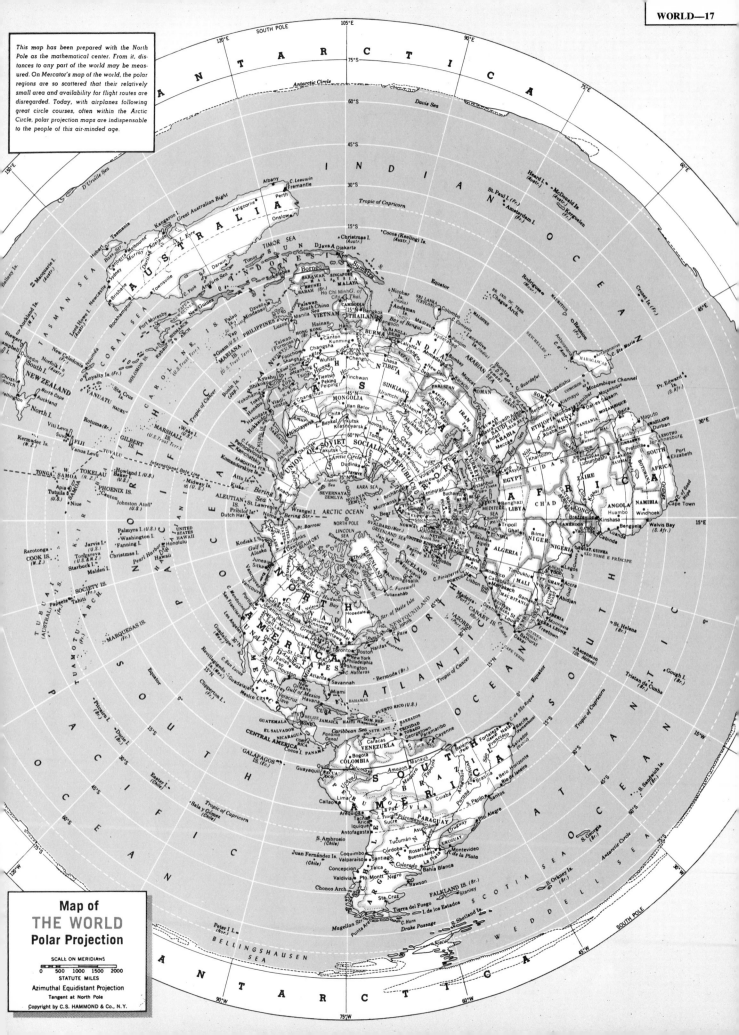

This map has been prepared with the North Pole as the mathematical center. From it, distances to any part of the world may be measured. On Mercator's map of the world, the polar regions are so scattered that their relatively small area and availability for flight routes are disregarded. Today, with airplanes following great circle courses, often within the Arctic Circle, polar projection maps are indispensable to the people of this air-minded age.

Map of
THE WORLD
Polar Projection

SCALE ON MERIDIANS

0   500   1000   1500   2000
STATUTE MILES

Azimuthal Equidistant Projection
Tangent at North Pole
Copyright by C.S. HAMMOND & Co., N.Y.

THE WORLD'S CONTINENTS BY SIZE

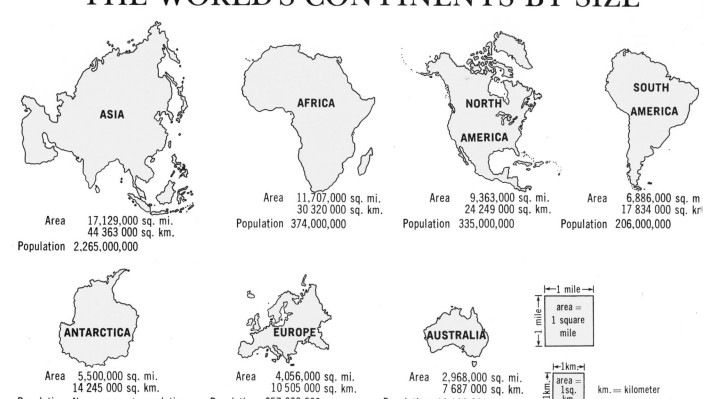

ASIA
Area   17,129,000 sq. mi.
      44 363 000 sq. km.
Population   2,265,000,000

AFRICA
Area   11,707,000 sq. mi.
      30 320 000 sq. km.
Population   374,000,000

NORTH AMERICA
Area   9,363,000 sq. mi.
      24 249 000 sq. km.
Population   335,000,000

SOUTH AMERICA
Area   6,886,000 sq. m
      17 834 000 sq. kr
Population   206,000,000

ANTARCTICA
Area   5,500,000 sq. mi.
      14 245 000 sq. km.
Population   No permanent population

EUROPE
Area   4,056,000 sq. mi.
      10 505 000 sq. km.
Population   657,000,000

AUSTRALIA
Area   2,968,000 sq. mi.
      7 687 000 sq. km.
Population   13,132,000

← 1 mile →
area = 1 square mile

← 1km →
area = 1sq. km.

km. = kilometer
sq. km. = square kilomete

WORLD TIME ZONES

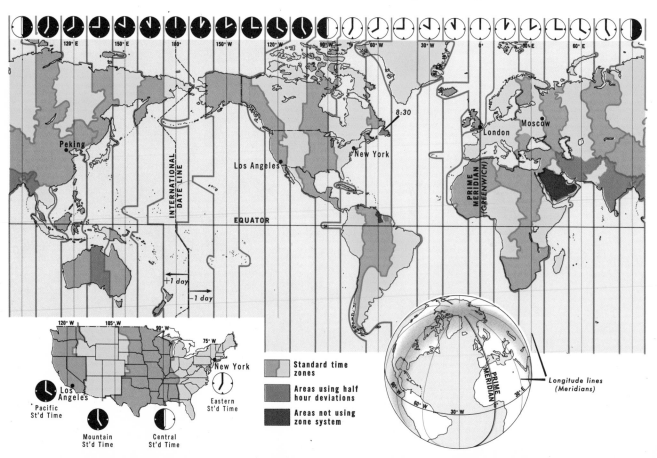

Peking
International Date Line
Los Angeles
New York
8:30
London
Moscow
Prime Meridian (Greenwich)
Equator
+1 day
−1 day

120° W   105° W   90° W   75° W
New York
Los Angeles
Pacific St'd Time
Mountain St'd Time
Central St'd Time
Eastern St'd Time

Standard time zones
Areas using half hour deviations
Areas not using zone system

Longitude lines (Meridians)
PRIME MERIDIAN

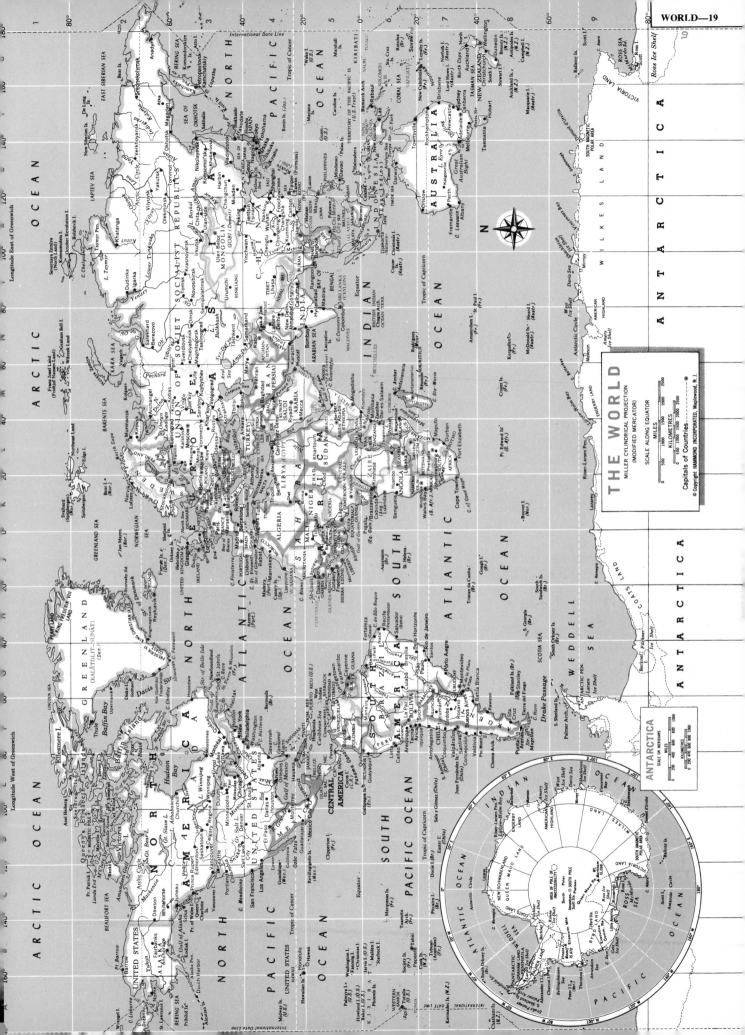

THE WORLD

MILLER CYLINDRICAL PROJECTION
(MODIFIED MERCATOR)

SCALE ALONG EQUATOR

MILES

500	1000	1500	2000	2500	

KILOMETRES

500	1000	1500	2000	2500	

Capitals of Countries............●

© Copyright HAMMOND INCORPORATED, Maplewood, N.J.

ANTARCTICA

SCALE ON MERIDIANS

MILES

200	400	600	800 1000

KILOMETRES

200 400	600	800	1000

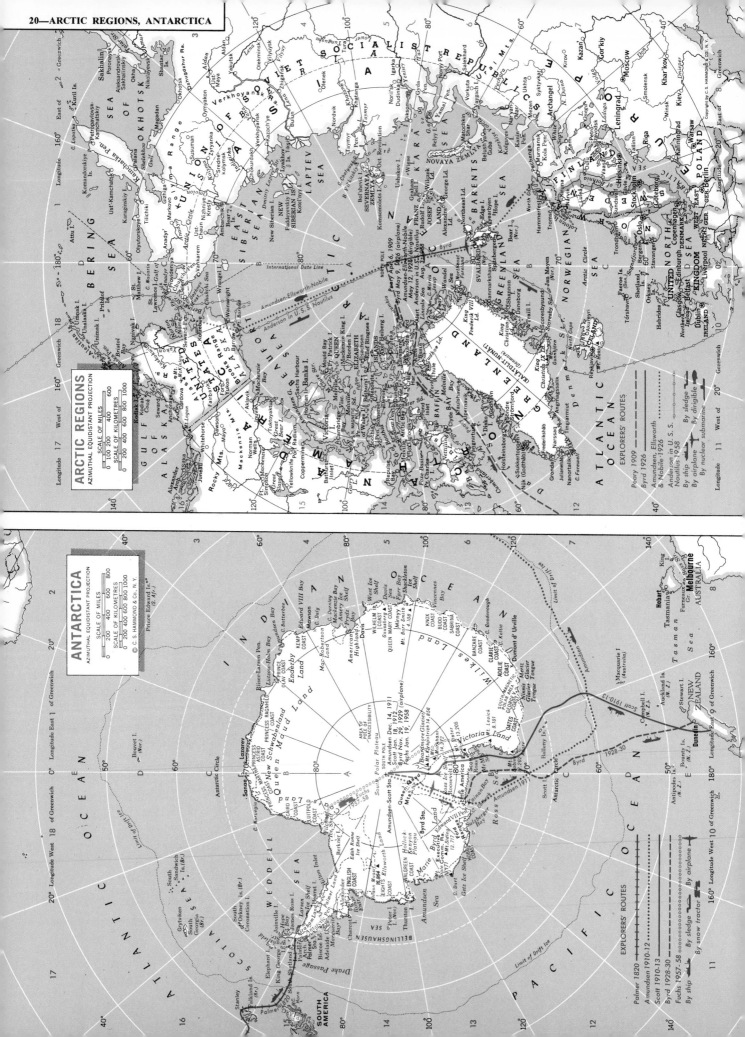

ARCTIC REGIONS
AZIMUTHAL EQUIDISTANT PROJECTION

SCALE OF MILES
0 100 200 400 600

SCALE OF KILOMETRES
0 200 400 600 800 1000

© C. S. HAMMOND & CO., N.Y.

EXPLORERS' ROUTES

Peary 1909
Byrd 1926
Amundsen, Ellsworth & Nobile 1926
Anderson in U.S.S. Nautilus 1958

By sledge
By airplane
By dirigible
By nuclear submarine

ANTARCTICA
AZIMUTHAL EQUIDISTANT PROJECTION

SCALE OF MILES
0 200 400 600 800

SCALE OF KILOMETRES
0 200 400 600 800 1000

© C. S. HAMMOND & CO., N.Y.

EXPLORERS' ROUTES

Palmer 1820
Amundsen 1910-12
Scott 1910-13
Byrd 1928-30
Fuchs 1957-58

By ship
By sledge
By airplane
By snow tractor

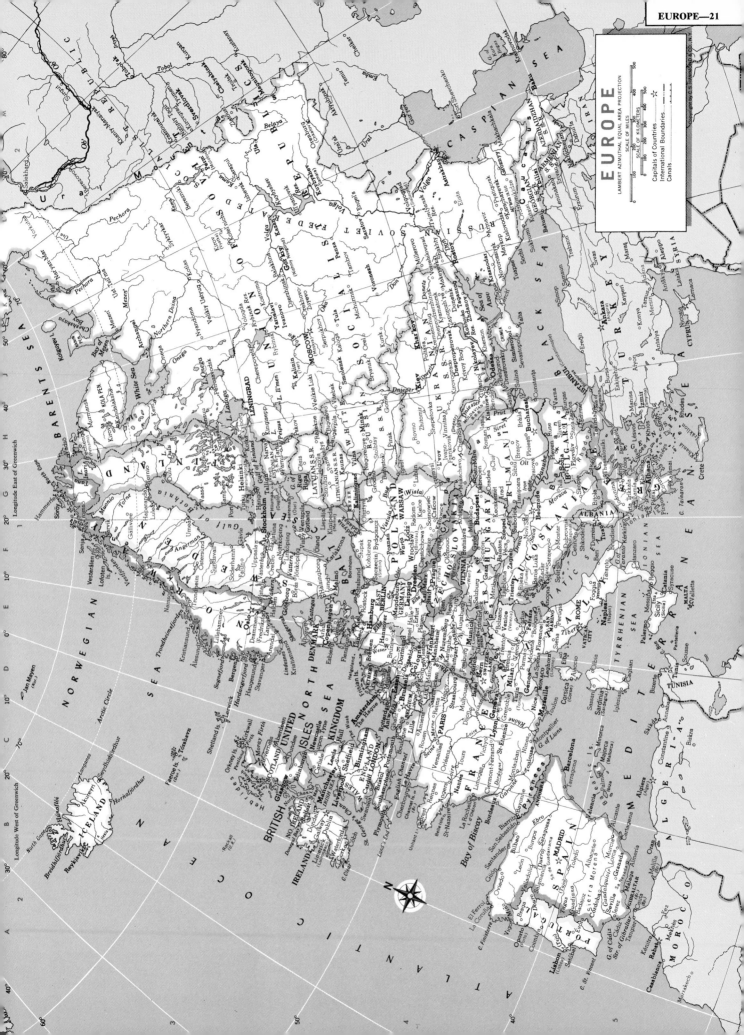

EUROPE

LAMBERT AZIMUTHAL EQUAL AREA PROJECTION

SCALE OF MILES

SCALE OF KILOMETERS

Capitals of Countries .... ☆
International Boundaries ....
Canals ....

Copyright by C.S. HAMMOND & CO., N.Y.

UNITED KINGDOM and IRELAND
BONNE PROJECTION

SCALE OF MILES
0 10 20 40 60 80

SCALE OF KILOMETRES

Capitals of Countries..........★
International Boundaries..........
Other Boundaries..........
Canals..........

SHETLAND ISLANDS
Same scale as main map.

SHETLAND ISLANDS
ORKNEY ISLANDS

IRISH SEA
NORTH SEA
ATLANTIC OCEAN
ENGLISH CHANNEL

NORTHERN IRELAND
IRELAND
SCOTLAND
ENGLAND
WALES

GREATER LONDON

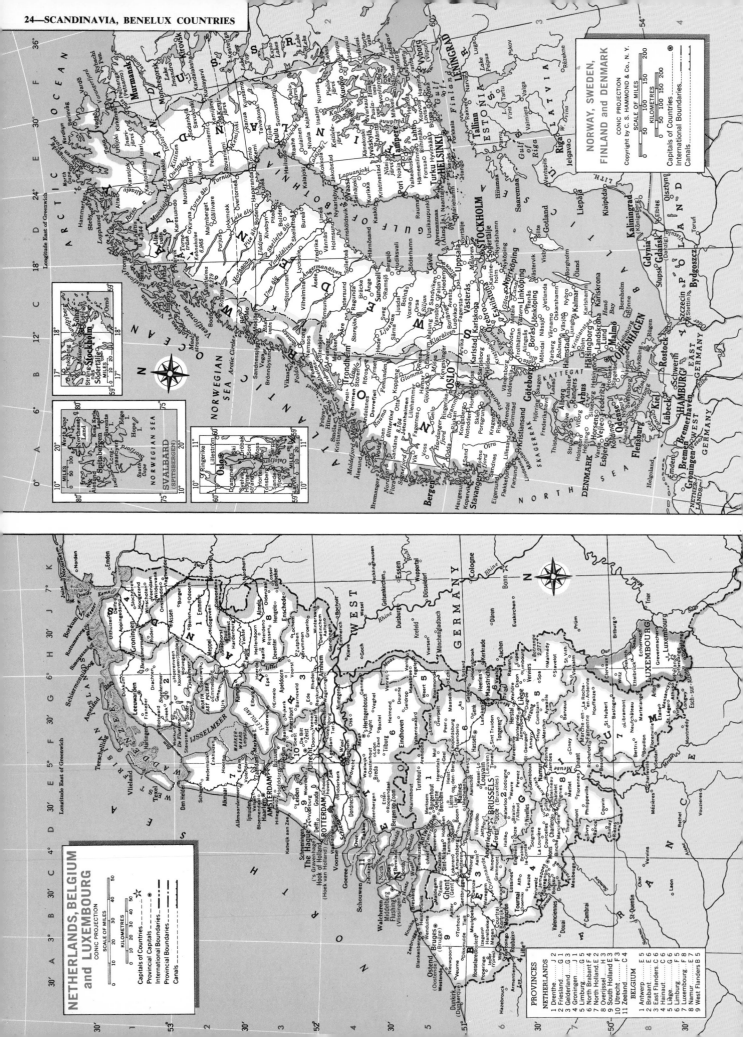

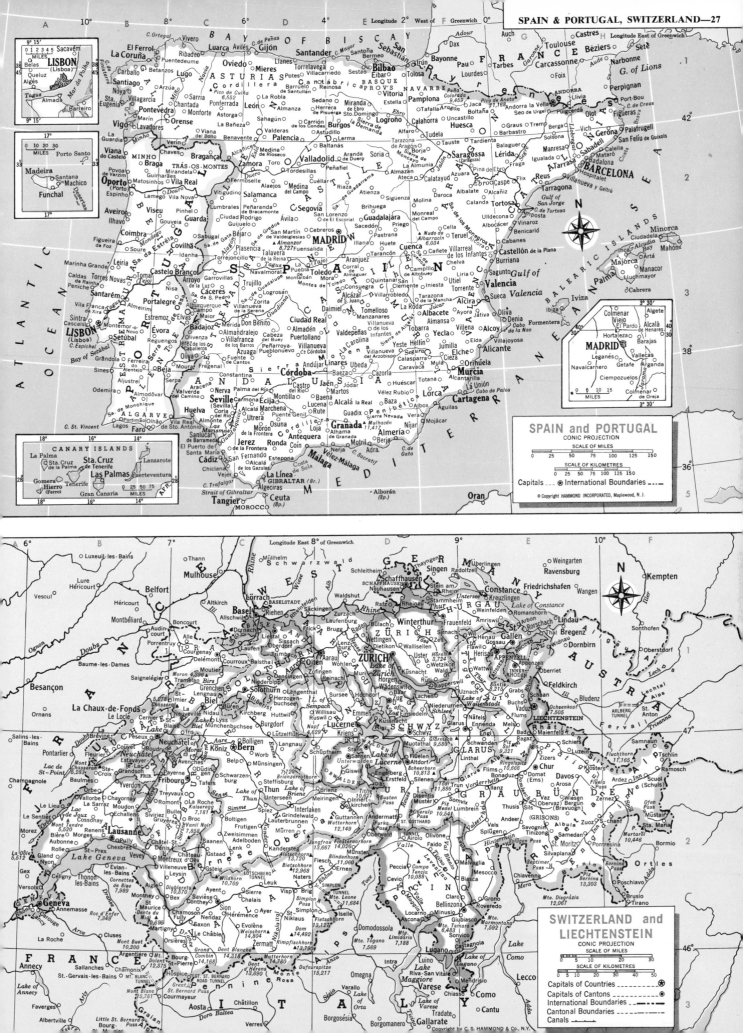

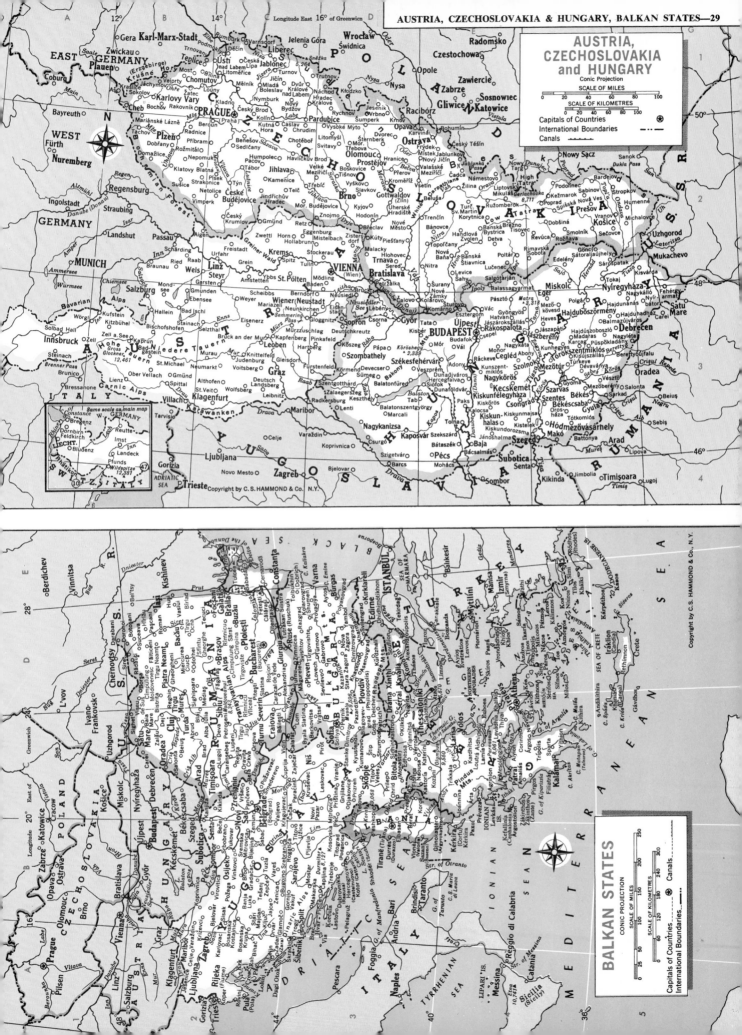

UNION OF SOVIET SOCIALIST REPUBLICS

CONIC PROJECTION

SCALE OF MILES

SCALE OF KILOMETRES

Capitals | Boundaries
Capitals:
● National
⊛ Union Republic
⊙ A.S.S.R.
○ Autonomous Oblast
◉ Autonomous Okrug

Boundaries:
National
Union Republic
A.S.S.R.
Autonomous Oblast
Autonomous Okrug

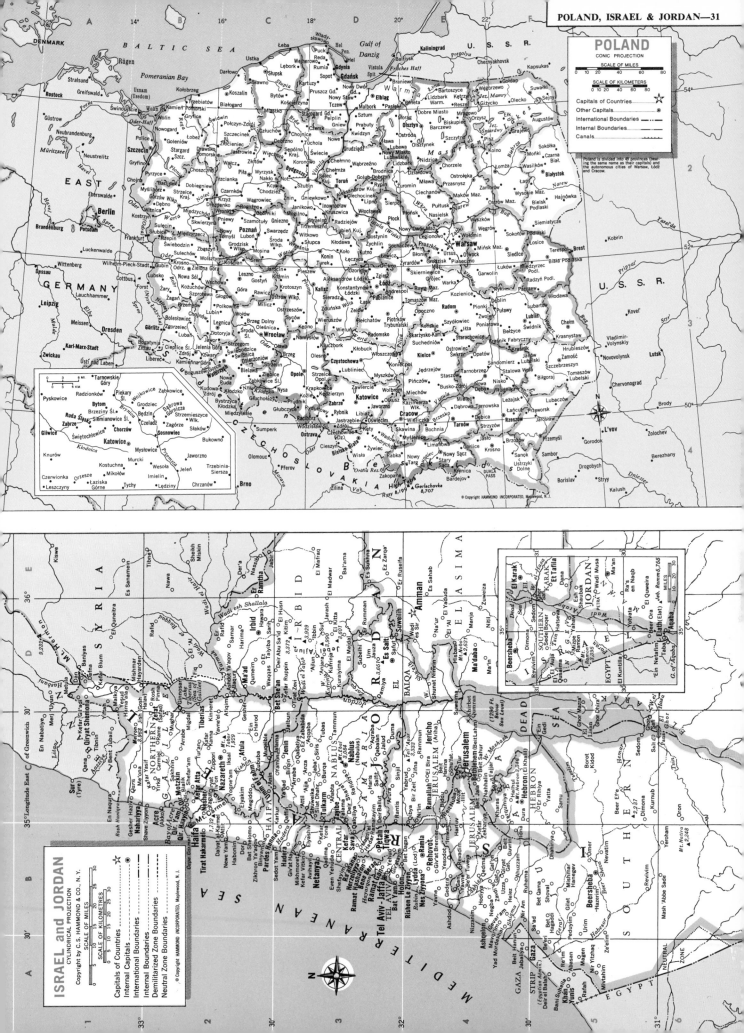

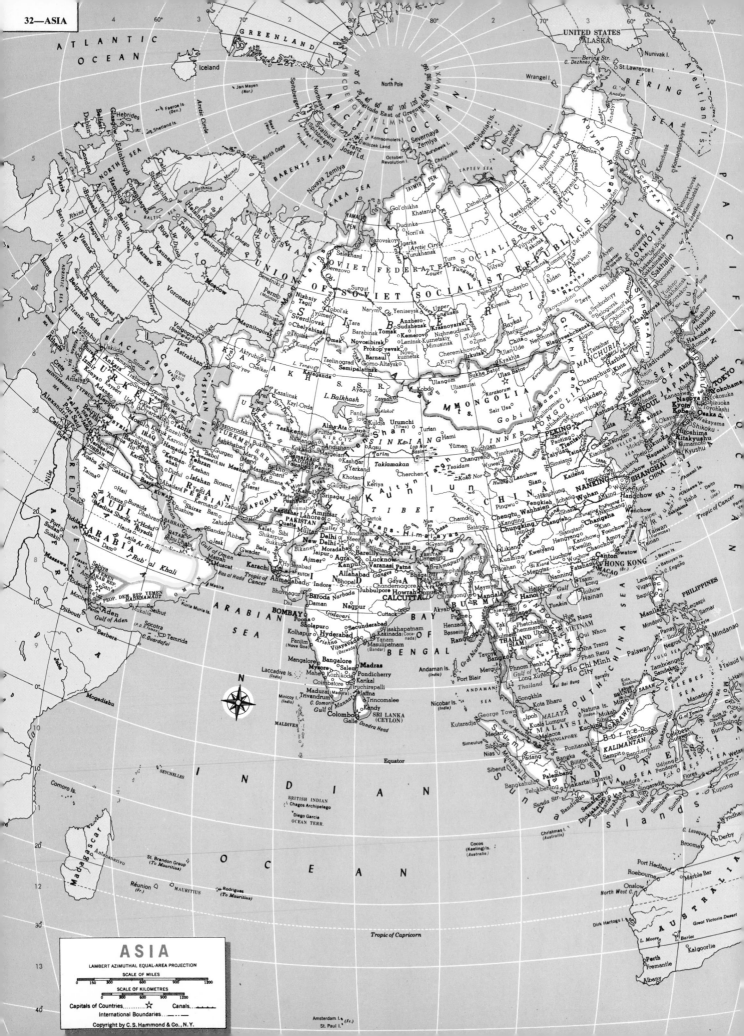

NEAR and MIDDLE EAST

CONIC PROJECTION
SCALE OF KILOMETRES

SCALE OF MILES

Capitals of Countries .... ☆
International Boundaries ....

Copyright by C. S. Hammond & Co., N.Y.

INDIAN SUBCONTINENT
and AFGHANISTAN

CONIC PROJECTION

SCALE OF MILES

SCALE OF KILOMETRES

Capitals of Countries ....☆
Provincial and State Capitals ......◉
International Boundaries ....
Provincial and State Boundaries ....
Canals ....

A R A B I A N

S E A

B A Y

O F

B E N G A L

I N D I A N   O C E A N

MALDIVES

SRI LANKA
(CEYLON)

ANDAMAN AND NICOBAR ISLANDS

CHINA and MONGOLIA

CONIC PROJECTION

SCALE OF MILES

SCALE OF KILOMETRES

Capitals of Countries..........☆ International Boundaries.........
Provincial Capitals.............● Provincial Boundaries...........
Canals............................ Walls...........................

© Copyright HAMMOND INCORPORATED, Maplewood, N.J.

*Wuhan municipality consists of
Hankow, Hanyang and Wuchang

AFRICA

LAMBERT AZIMUTHAL EQUAL-AREA PROJECTION

SCALE OF MILES
0   100   200   400   600

SCALE OF KILOMETRES
0 100 200   400   600

Copyright by C.S. Hammond & Co., N.Y.

Capitals of Countries .... ⭐
Other Capitals .... ◉
International Boundaries .... ▬▬▬
Internal Boundaries .... ▪▪▪
Canals .... Wells ◦

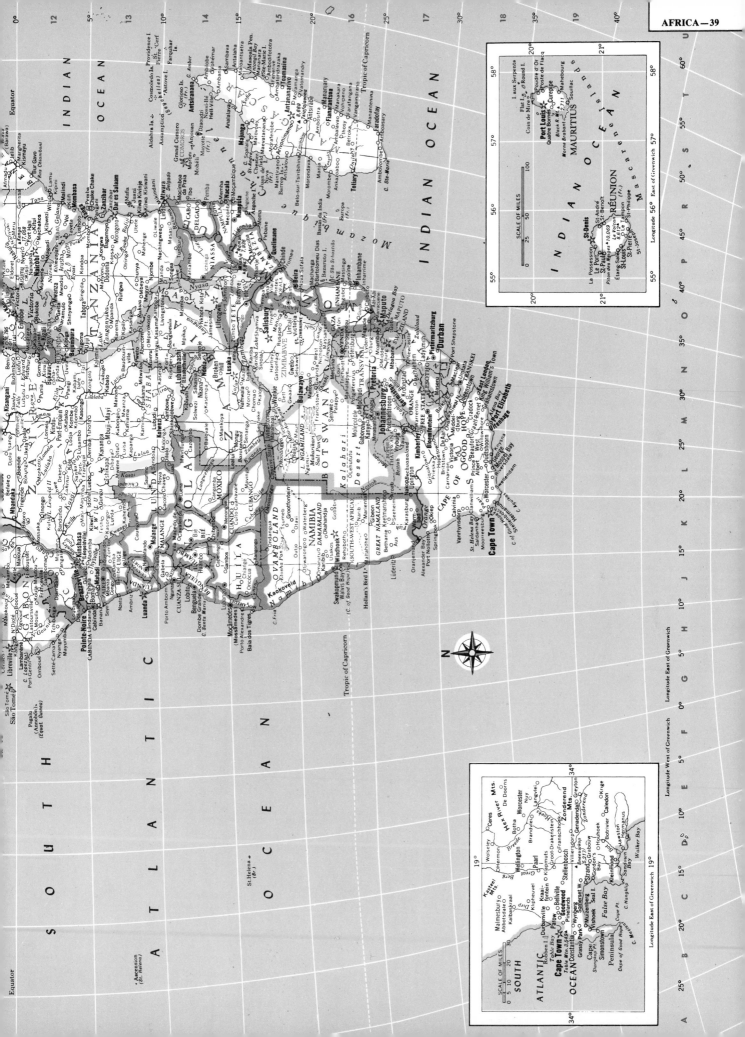

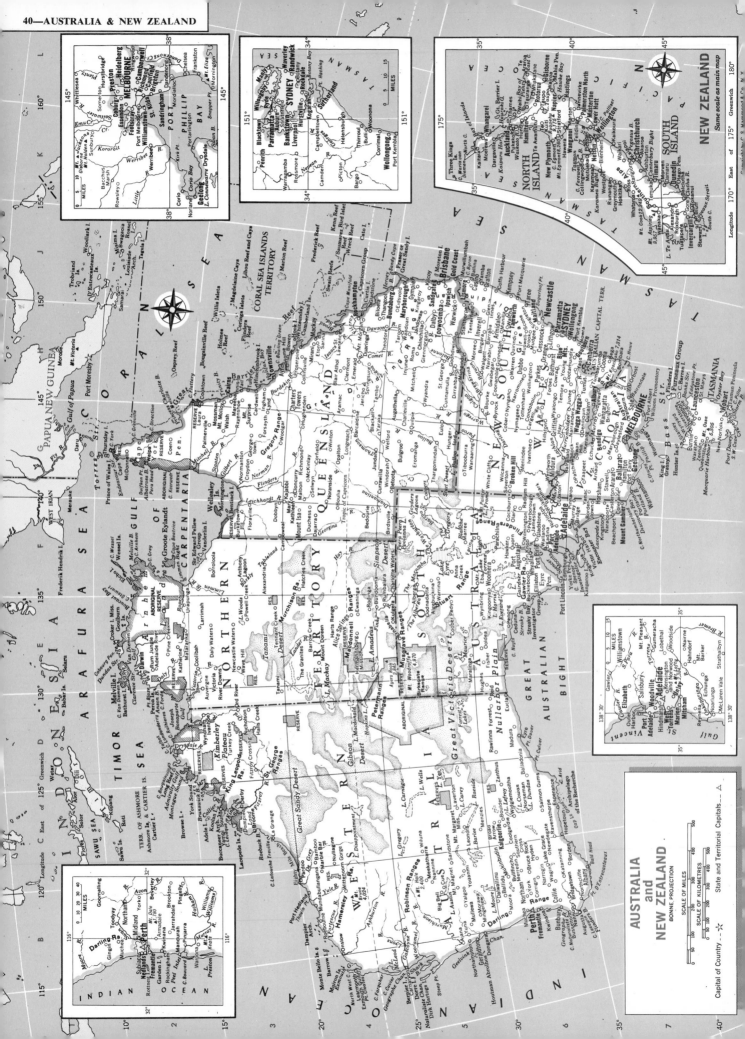

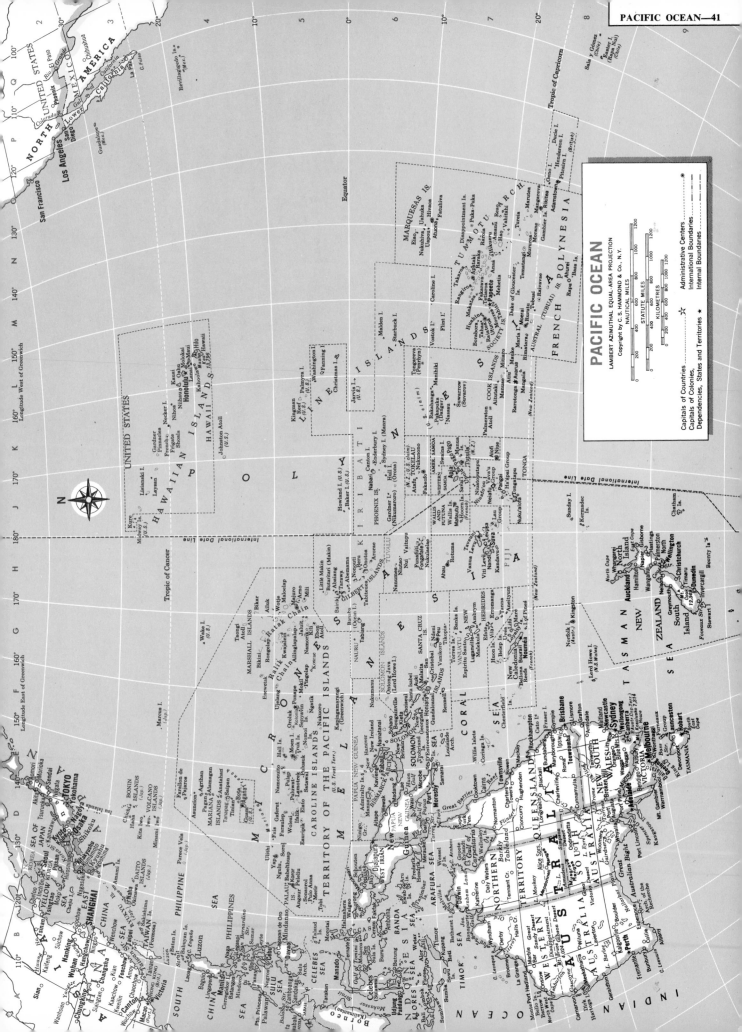

PACIFIC OCEAN

LAMBERT AZIMUTHAL EQUAL-AREA PROJECTION
Copyright by C. S. Hammond & Co., N.Y.

NAUTICAL MILES
STATUTE MILES
KILOMETRES

Capitals of Countries ....
Capitals of Colonies,
Dependencies, States and Territories .... ★

Administrative Centers ....
International Boundaries ....
Internal Boundaries ....

NORTH AMERICA

LAMBERT AZIMUTHAL EQUAL-AREA PROJECTION

SCALE OF MILES

0 100 200   400   600   800

SCALE OF KILOMETRES

0   200   400   600   800

Capitals of Countries _____ ☆
International Boundaries _____ — — —
Other Boundaries _____ — · — ·
Canals _____ — · · — · ·

© C.S. HAMMOND & CO., N.Y.

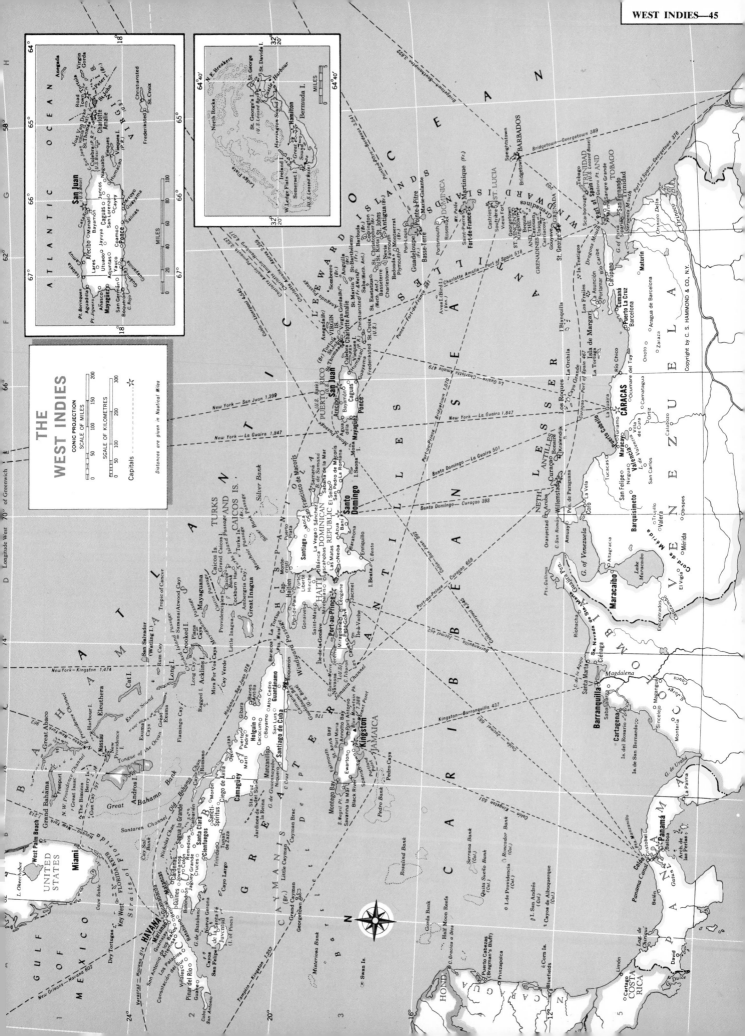

THE WEST INDIES

CONIC PROJECTION

SCALE OF MILES

SCALE OF KILOMETRES

Capitals .... ☆

Distances are given in Nautical Miles

San Juan

ATLANTIC OCEAN

N.E. Breakers

North Rocks

St. George's I.
(U.S. Naval Base)

Castle Harbour

Hamilton

Bermuda I.

W. Ledge Flats

Somerset I.

ATLANTIC OCEAN

Copyright by C. S. HAMMOND & CO., N.Y.

LEEWARD ISLANDS

BARBADOS

Bridgetown—Georgetown 389

TRINIDAD

TOBAGO

WINDWARD ISLANDS

LESSER ANTILLES

NETH. ANTILLES

CARACAS

VENEZUELA

Maracaibo

Lake Maracaibo

Barranquilla

Cartagena

PANAMA

Panama Canal

COSTA RICA

NICARAGUA

HONDURAS

CARIBBEAN SEA

JAMAICA

Kingston

New York — Kingston 1,474

Kingston — Barranquilla 437

HAITI

DOMINICAN REPUBLIC

Port-au-Prince

Santo Domingo

Santo Domingo — La Guaira 501

Santo Domingo — Curaçao 393

PUERTO RICO

San Juan

New York — San Juan 1,399

New York — La Guaira 1,847

BAHAMA ISLANDS

CUBA

HAVANA

UNITED STATES

Miami

GULF OF MEXICO

Straits of Florida

TURKS AND CAICOS IS.

Tropic of Cancer

Silver Bank

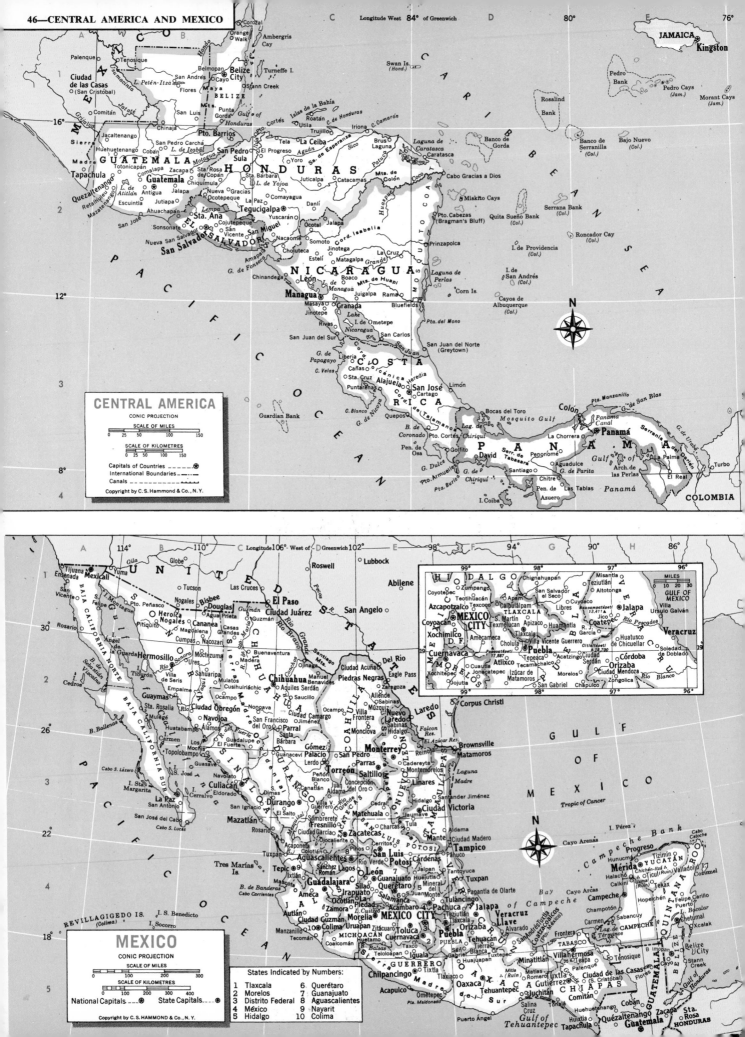

CANADA

CONIC PROJECTION

SCALE OF MILES

SCALE OF KILOMETRES

Capitals of Countries .... ★
Provincial Capitals ....
International Boundaries ....
Provincial Boundaries ....
Canals ....

Copyright by C.S. HAMMOND & Co., N.Y.

QUEEN ELIZABETH ISLANDS

Scale of Miles

POPULATION DISTRIBUTION

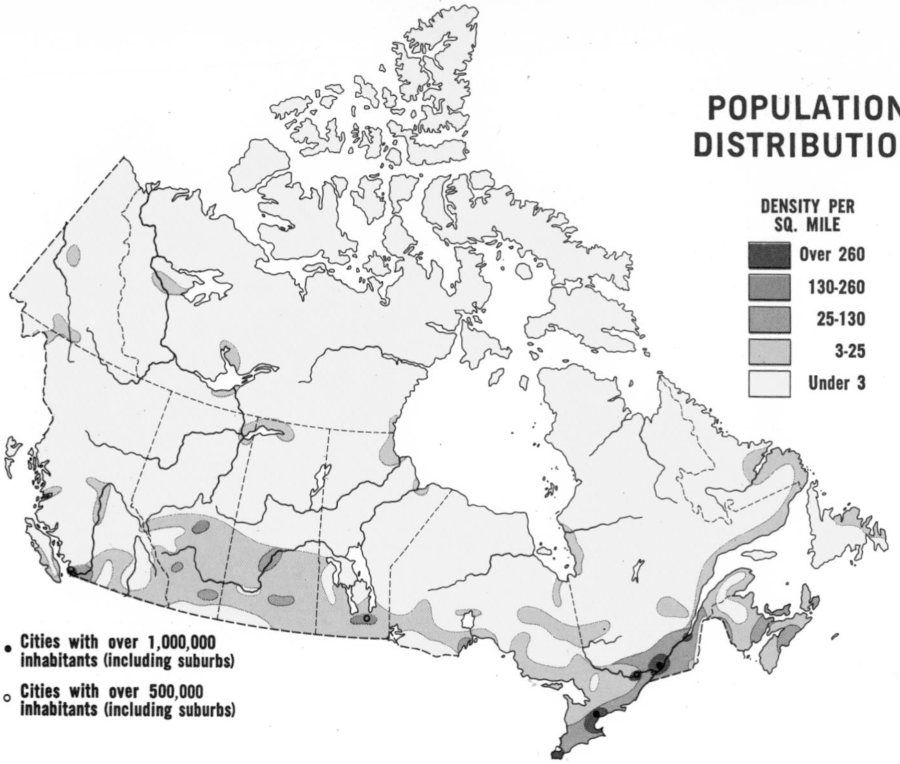

DENSITY PER SQ. MILE
Over 260
130-260
25-130
3-25
Under 3
Cities with over 1,000,000 inhabitants (including suburbs)
Cities with over 500,000 inhabitants (including suburbs)

AGRICULTURE, INDUSTRY and RESOURCES

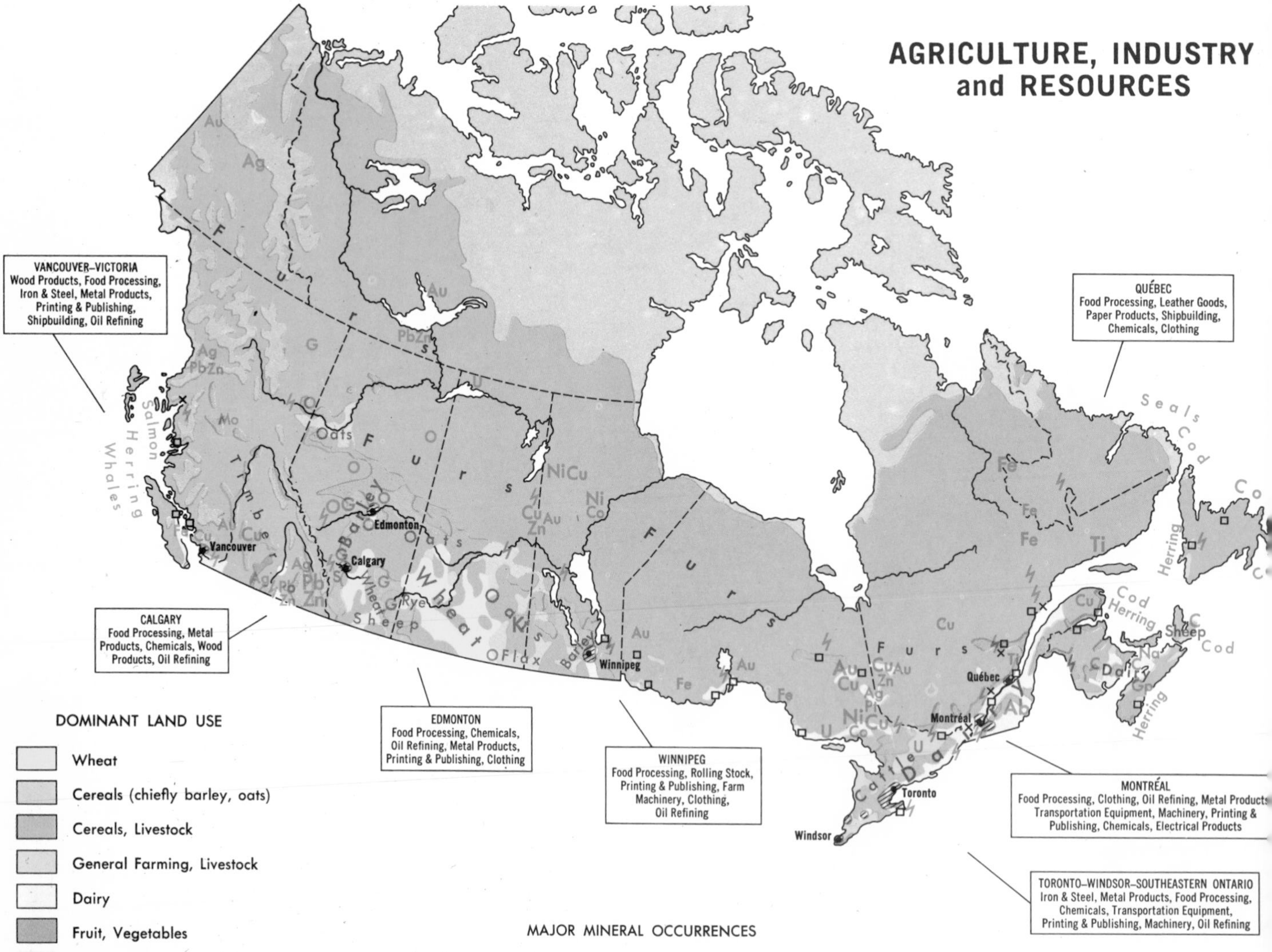

VANCOUVER–VICTORIA
Wood Products, Food Processing, Iron & Steel, Metal Products, Printing & Publishing, Shipbuilding, Oil Refining
CALGARY
Food Processing, Metal Products, Chemicals, Wood Products, Oil Refining
EDMONTON
Food Processing, Chemicals, Oil Refining, Metal Products, Printing & Publishing, Clothing
WINNIPEG
Food Processing, Rolling Stock, Printing & Publishing, Farm Machinery, Clothing, Oil Refining
QUÉBEC
Food Processing, Leather Goods, Paper Products, Shipbuilding, Chemicals, Clothing
MONTRÉAL
Food Processing, Clothing, Oil Refining, Metal Products, Transportation Equipment, Machinery, Printing & Publishing, Chemicals, Electrical Products
TORONTO–WINDSOR–SOUTHEASTERN ONTARIO
Iron & Steel, Metal Products, Food Processing, Chemicals, Transportation Equipment, Printing & Publishing, Machinery, Oil Refining
Vancouver
Edmonton
Calgary
Winnipeg
Québec
Montréal
Toronto
Windsor
Furs
Timber
Salmon
Herring
Whales
Seals
Cod
Oats
Barley
Wheat
Rye
Sheep
Flax

DOMINANT LAND USE

- Wheat
- Cereals (chiefly barley, oats)
- Cereals, Livestock
- General Farming, Livestock
- Dairy
- Fruit, Vegetables
- Pasture Livestock
- Range Livestock
- Forests
- Nonagricultural Land

MAJOR MINERAL OCCURRENCES

Ab Asbestos	Cu Copper	Mo Molybdenum	Pt Platinum
Ag Silver	Fe Iron Ore	Na Salt	S Sulfur
Au Gold	G Natural Gas	Ni Nickel	Ti Titanium
C Coal	Gp Gypsum	O Petroleum	U Uranium
Co Cobalt	K Potash	Pb Lead	Zn Zinc

- Water Power
- Major Industrial Areas
- Major Pulp & Paper Mills
- × Aluminum Smelters

VEGETATION

MID-LATITUDE FOREST

Coniferous Forest

Broadleaf Forest

Mixed Coniferous and Broadleaf Forest

MID-LATITUDE GRASSLAND

Short Grass (Steppe)

Tall Grass (Prairie)

DESERT AND DESERT SHRUB

TUNDRA AND ALPINE

PERMANENT ICE

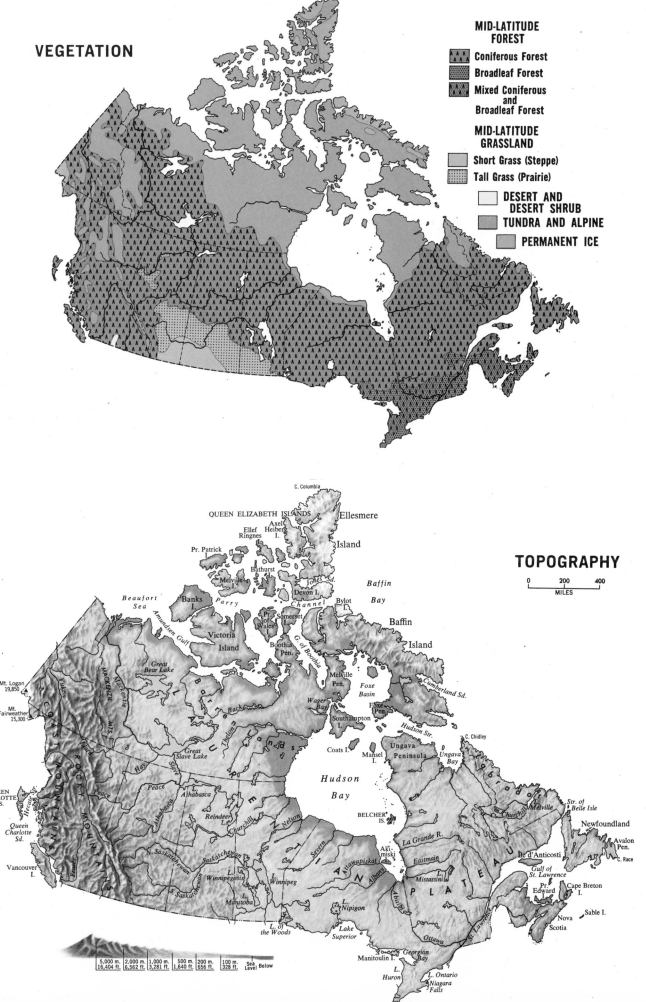

TOPOGRAPHY

0   200   400
MILES

C. Columbia

QUEEN ELIZABETH ISLANDS

Ellesmere

Ellef Ringnes I.

Axel Heiberg I.

Pr. Patrick

Island

Bathurst

Melville I.

Banks I.

Parry

Jones Sd.

Devon I.

Baffin Bay

Beaufort Sea

Amundsen Gulf

Pr. of Wales I.

Somerset I.

Bylot

Baffin

Victoria Island

Boothia Pen.

G. of Boothia

Island

Mt. Logan 19,850

Mackenzie

Great Bear Lake

Back

Thelon

Melville Pen.

Foxe Basin

Cumberland Sd.

Mt. Fairweather 15,300

Wager Bay

Foxe Pen.

Hudson Str.

C. Chidley

Great Slave Lake

Southampton I.

Coats I.

Mansel I.

Ungava Peninsula

Ungava Bay

QUEEN CHARLOTTE IS.

Hecate Str.

Peace

Slave

Athabasca

Peace

Reindeer Lk.

Churchill

Nelson

Hudson Bay

BELCHER IS.

Aki-miski I.

Churchill

Melville

Str. of Belle Isle

Queen Charlotte Sd.

N. Saskatchewan

Severn

Winnipegosis

Winnipeg

Manitoba

Nipigon

Lake Superior

Albany

Attawapiskat

La Grande R.

Eastmain

Mistassini

Newfoundland

Avalon Pen.

C. Race

Ile d'Anticosti

Gulf of St. Lawrence

Cape Breton I.

Pr. Edward

Sable I.

Nova Scotia

Vancouver I.

Fraser

S. Saskatchewan

L. of the Woods

Lake Superior

Manitoulin I.

Georgian Bay

Ottawa

L. Huron

L. Ontario

Niagara Falls

St. Lawrence

PLATEAU

LABRADOR

ROCKY MOUNTAINS

COAST MOUNTAINS

LAURENTIAN

5,000 m. 16,404 ft. | 2,000 m. 6,562 ft. | 1,000 m. 3,281 ft. | 500 m. 1,640 ft. | 200 m. 656 ft. | 100 m. 328 ft. | Sea Level | Below

NEWFOUNDLAND
INCLUDING LABRADOR

SCALE

Capitals of Provinces .... ✦
Provincial Boundaries .... ▬ ▪ ▬ ▪
Provincial Boundary according to
Imperial Privy Council decision, 1927 .... ▬ ▬ ▬

© Copyright by C. S. HAMMOND & Co.

Longitude West of Greenwich

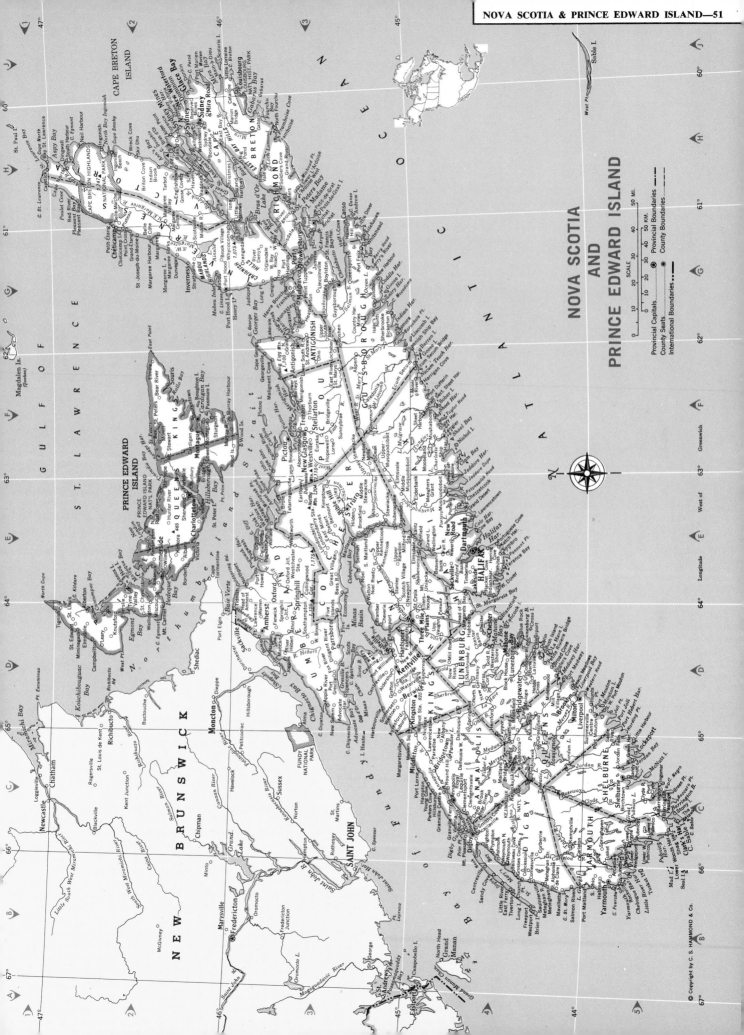

NOVA SCOTIA
AND
PRINCE EDWARD ISLAND

SCALE

Provincial Capitals..........⊛
County Seats.................⊙
Provincial Boundaries........
County Boundaries............
International Boundaries......

© Copyright by C. S. HAMMOND & Co.

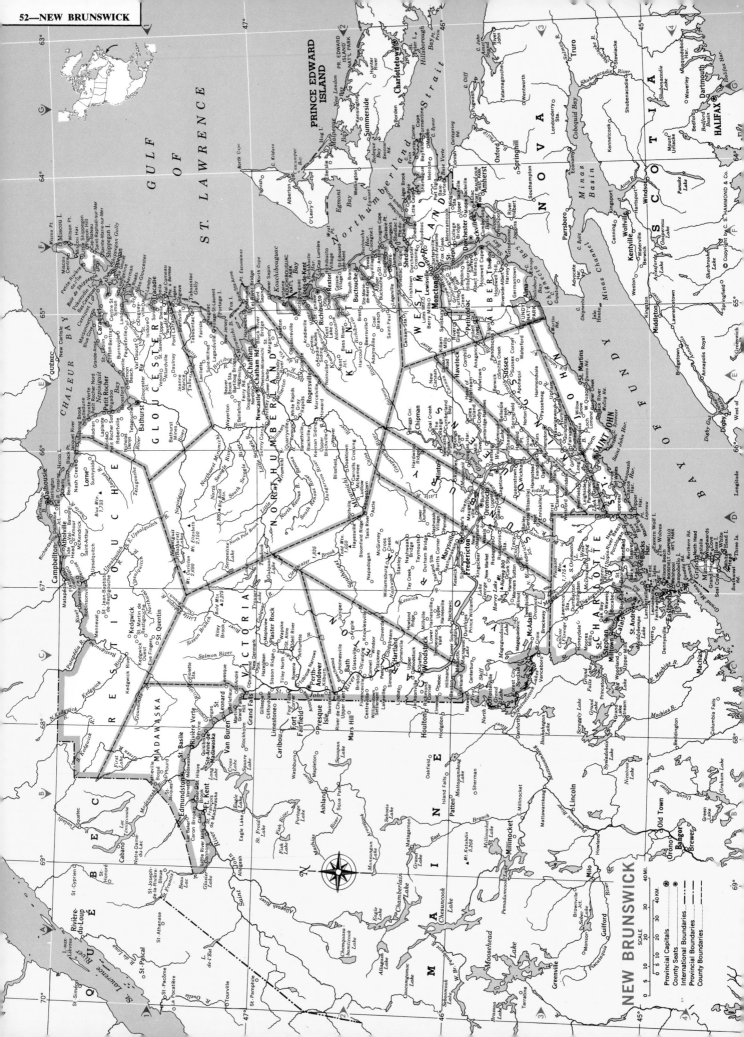

NEW BRUNSWICK

SCALE
0 5 10 20 30 40 MI.
0 5 10 20 30 40 KM.

Provincial Capitals ⊛
County Seats ⊙
International Boundaries
Provincial Boundaries
County Boundaries

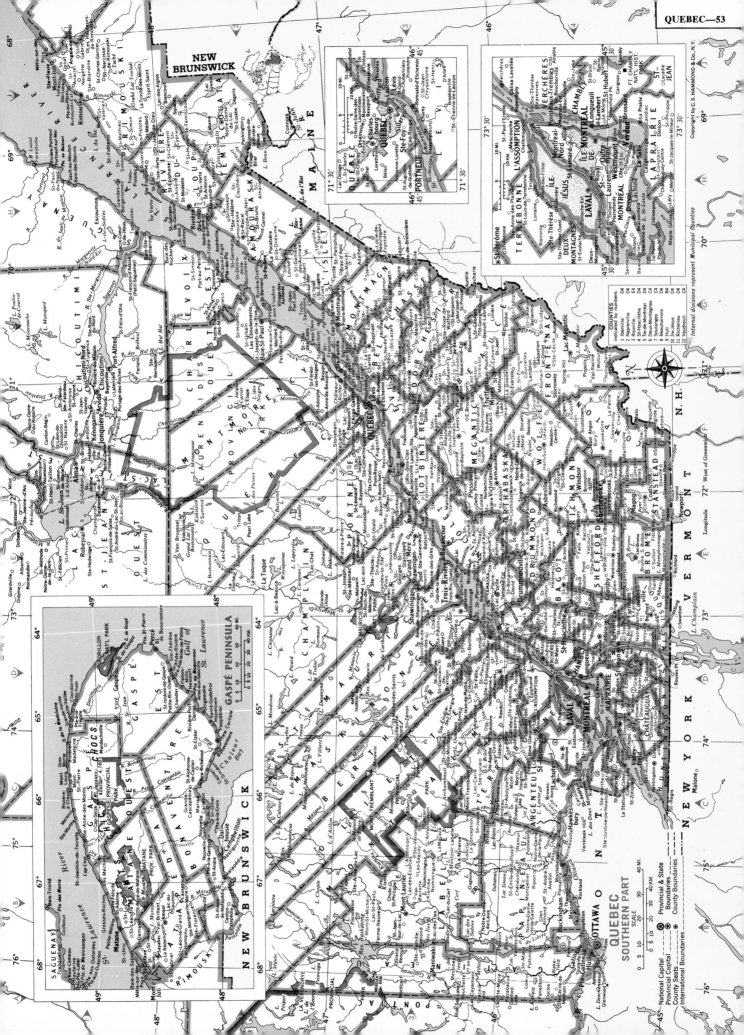

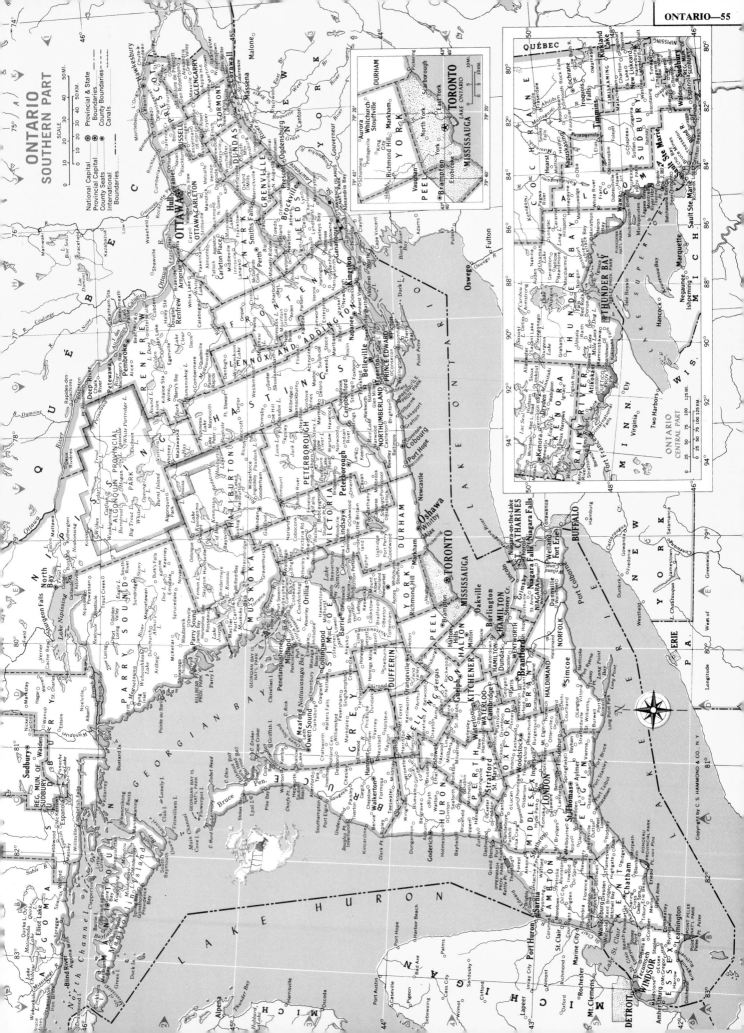

ONTARIO
SOUTHERN PART

SCALE

National Capital ⊛
Provincial Capital ⊛
County Seats ⊙
Boundaries

Provincial & State Boundaries
County Boundaries
Canals
International Boundaries

QUÉBEC

ONTARIO CENTRAL PART

TORONTO
DURHAM
YORK
PEEL

Copyright by C. S. HAMMOND & CO., N.Y.

MANITOBA
NORTHERN PART

MANITOBA
SOUTHERN PART

SCALE

Provincial Capital ⊛
International Boundaries
Provincial Boundaries

© C.S. HAMMOND & CO., N.Y.

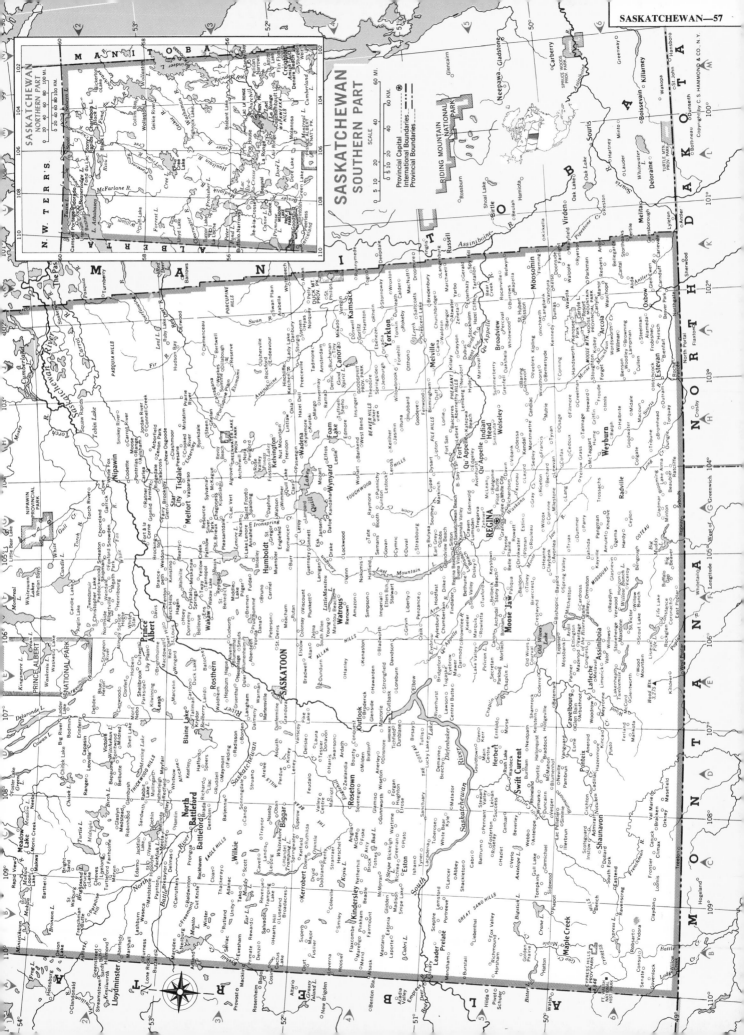

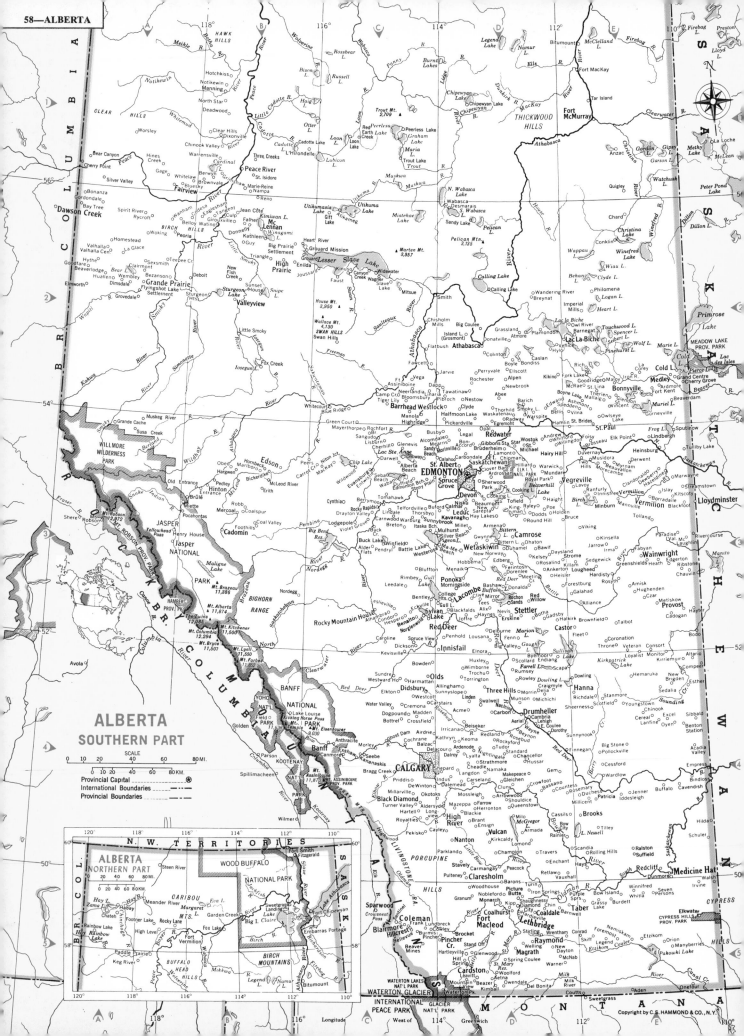

ALBERTA
SOUTHERN PART

SCALE
0 10 20 40 60 80 MI.
0 10 20 40 60 80 KM.

Provincial Capital ⊛
International Boundaries − − −
Provincial Boundaries

ALBERTA
NORTHERN PART
0 20 40 60 80 MI.
0 20 40 60 80 KM.

N. W. TERRITORIES
WOOD BUFFALO
NATIONAL PARK
CARIBOU MTS.
BIRCH MOUNTAINS
BUFFALO HEAD HILLS

YUKON AND NORTHWEST TERRITORIES

SCALE

MI.					
0	50	100	200	300	
0	50 100	200	300 KM.		

⊛ Territorial Capitals
International Boundaries
Provincial & Territorial Boundaries
District Boundaries

All islands in Hudson and James Bays lie within the District of Keewatin.

Longitude West K of Greenwich

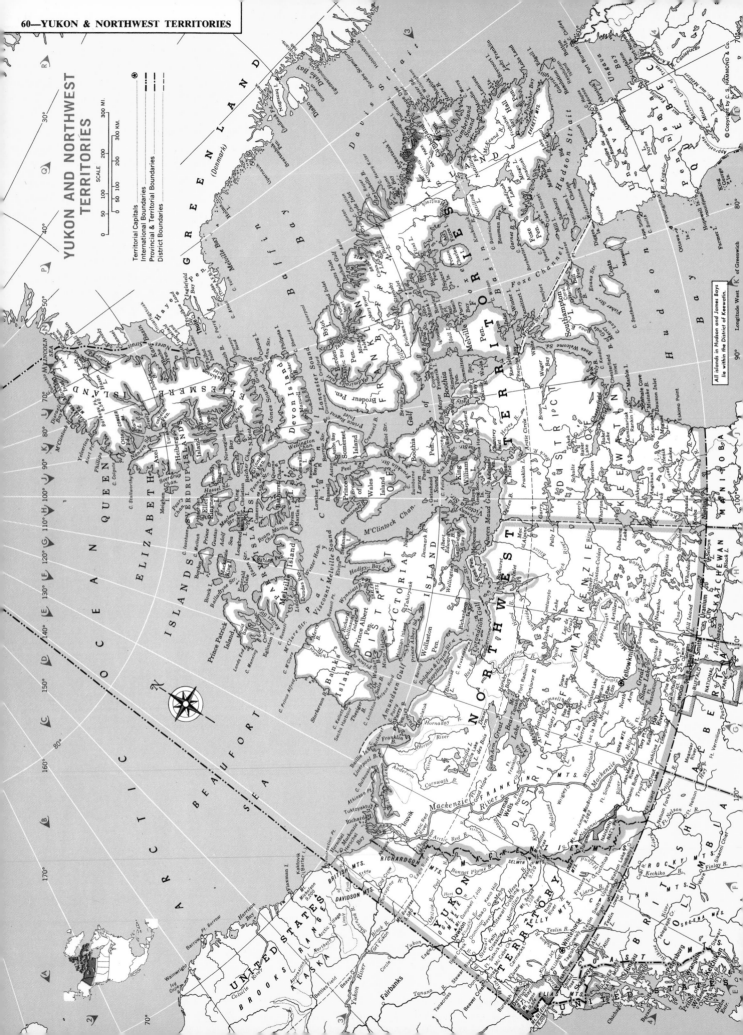

UNITED STATES

POLYCONIC PROJECTION

SCALE

| 0 50 100 | 200 | 300 | 400 MI. |

| 50 | 100 | 200 | 300 400 KM. |

Capitals of Countries ☆
State Capitals ▲
International Boundaries
State Boundaries

© C. S. HAMMOND & Co., N.Y.

POPULATION DISTRIBUTION

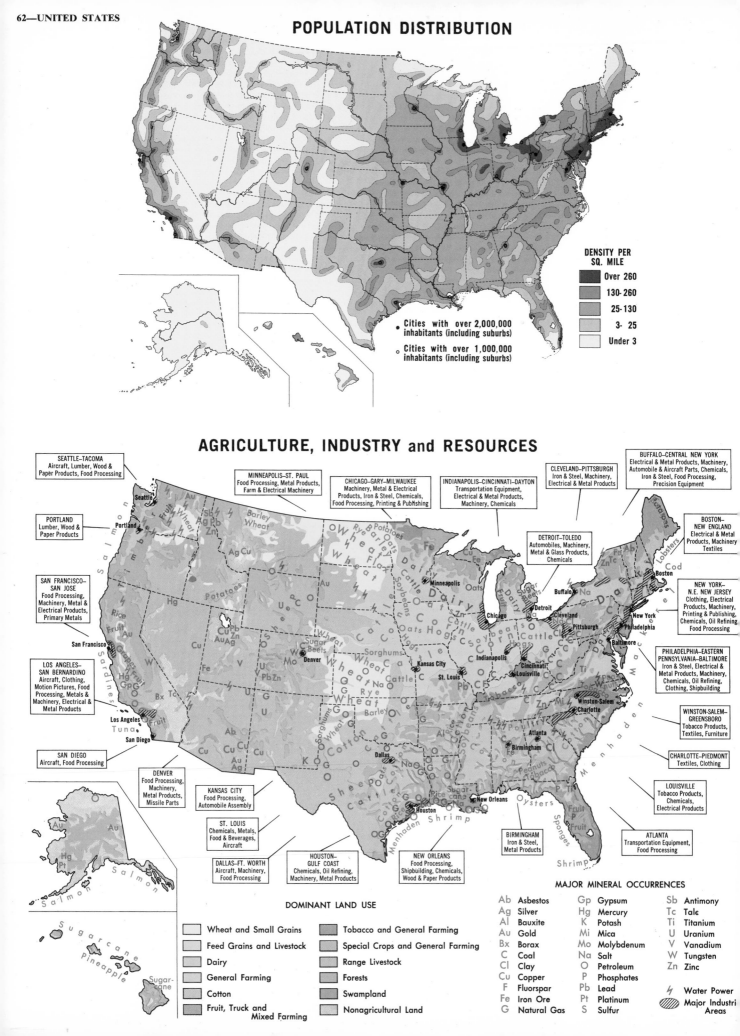

DENSITY PER SQ. MILE

Over 260
130- 260
25- 130
3- 25
Under 3

○ Cities with over 2,000,000 inhabitants (including suburbs)

○ Cities with over 1,000,000 inhabitants (including suburbs)

AGRICULTURE, INDUSTRY and RESOURCES

SEATTLE–TACOMA
Aircraft, Lumber, Wood & Paper Products, Food Processing

PORTLAND
Lumber, Wood & Paper Products

SAN FRANCISCO–SAN JOSE
Food Processing, Machinery, Metal & Electrical Products, Primary Metals

LOS ANGELES–SAN BERNARDINO
Aircraft, Clothing, Motion Pictures, Food Processing, Metals & Machinery, Electrical & Metal Products

SAN DIEGO
Aircraft, Food Processing

DENVER
Food Processing, Machinery, Metal Products, Missile Parts

KANSAS CITY
Food Processing, Automobile Assembly

ST. LOUIS
Chemicals, Metals, Food & Beverages, Aircraft

DALLAS–FT. WORTH
Aircraft, Machinery, Food Processing

HOUSTON–GULF COAST
Chemicals, Oil Refining, Machinery, Metal Products

NEW ORLEANS
Food Processing, Shipbuilding, Chemicals, Wood & Paper Products

MINNEAPOLIS–ST. PAUL
Food Processing, Metal Products, Farm & Electrical Machinery

CHICAGO–GARY–MILWAUKEE
Machinery, Metal & Electrical Products, Iron & Steel, Chemicals, Food Processing, Printing & Publishing

INDIANAPOLIS–CINCINNATI–DAYTON
Transportation Equipment, Electrical & Metal Products, Machinery, Chemicals

DETROIT–TOLEDO
Automobiles, Machinery, Metal & Glass Products, Chemicals

CLEVELAND–PITTSBURGH
Iron & Steel, Machinery, Electrical & Metal Products

BUFFALO–CENTRAL NEW YORK
Electrical & Metal Products, Machinery, Automobile & Aircraft Parts, Chemicals, Iron & Steel, Food Processing, Precision Equipment

BOSTON–NEW ENGLAND
Electrical & Metal Products, Machinery, Textiles

NEW YORK–N.E. NEW JERSEY
Clothing, Electrical Products, Machinery, Printing & Publishing, Chemicals, Oil Refining, Food Processing

PHILADELPHIA–EASTERN PENNSYLVANIA–BALTIMORE
Iron & Steel, Electrical & Metal Products, Machinery, Chemicals, Oil Refining, Clothing, Shipbuilding

WINSTON-SALEM–GREENSBORO
Tobacco Products, Textiles, Furniture

CHARLOTTE–PIEDMONT
Textiles, Clothing

LOUISVILLE
Tobacco Products, Chemicals, Electrical Products

ATLANTA
Transportation Equipment, Food Processing

BIRMINGHAM
Iron & Steel, Metal Products

DOMINANT LAND USE

Wheat and Small Grains
Feed Grains and Livestock
Dairy
General Farming
Cotton
Fruit, Truck and Mixed Farming
Tobacco and General Farming
Special Crops and General Farming
Range Livestock
Forests
Swampland
Nonagricultural Land

MAJOR MINERAL OCCURRENCES

Ab Asbestos	Gp Gypsum	Sb Antimony
Ag Silver	Hg Mercury	Tc Talc
Al Bauxite	K Potash	Ti Titanium
Au Gold	Mi Mica	U Uranium
Bx Borax	Mo Molybdenum	V Vanadium
C Coal	Na Salt	W Tungsten
Cl Clay	O Petroleum	Zn Zinc
Cu Copper	P Phosphates	
F Fluorspar	Pb Lead	⚡ Water Power
Fe Iron Ore	Pt Platinum	▨ Major Industrial Areas
G Natural Gas	S Sulfur	

VEGETATION

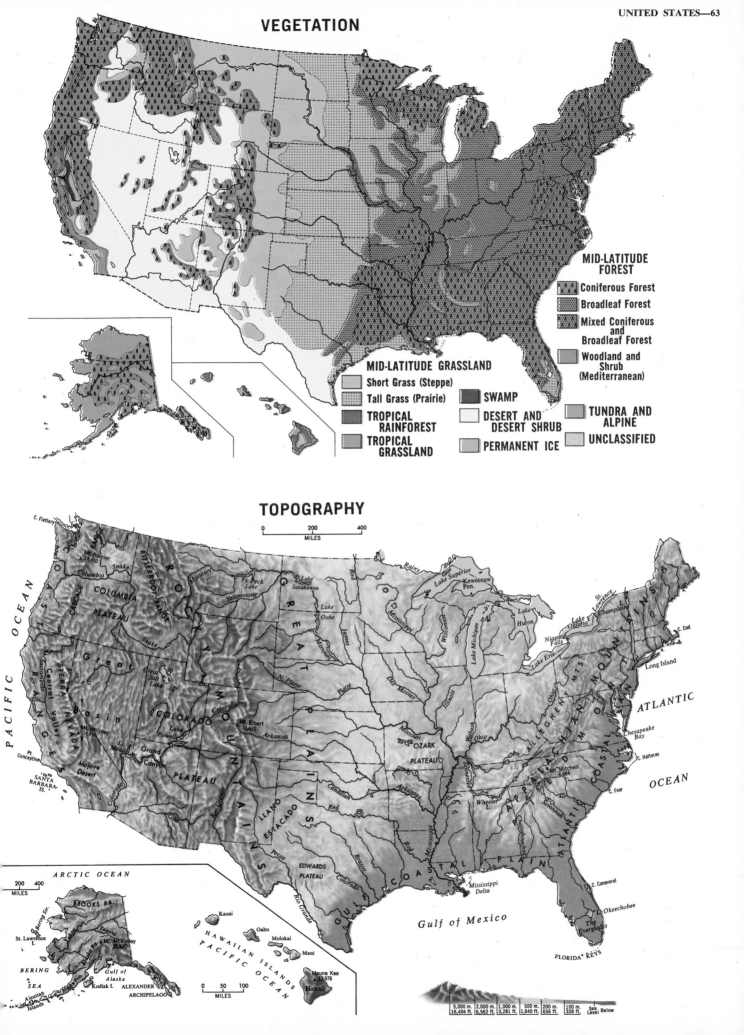

MID-LATITUDE FOREST

- ᴧᴧᴧ Coniferous Forest
- Broadleaf Forest
- Mixed Coniferous and Broadleaf Forest
- Woodland and Shrub (Mediterranean)

MID-LATITUDE GRASSLAND

- Short Grass (Steppe)
- Tall Grass (Prairie)
- **TROPICAL RAINFOREST**
- **TROPICAL GRASSLAND**

- SWAMP
- DESERT AND DESERT SHRUB
- PERMANENT ICE

- TUNDRA AND ALPINE
- UNCLASSIFIED

TOPOGRAPHY

0   200   400
MILES

200   400
MILES

0   50   100
MILES

5,000 m. | 2,000 m. | 1,000 m. | 500 m. | 200 m. | 100 m. | Sea
16,404 ft. | 6,562 ft. | 3,281 ft. | 1,640 ft. | 656 ft. | 328 ft. | Level | Below

TEMPERATURE

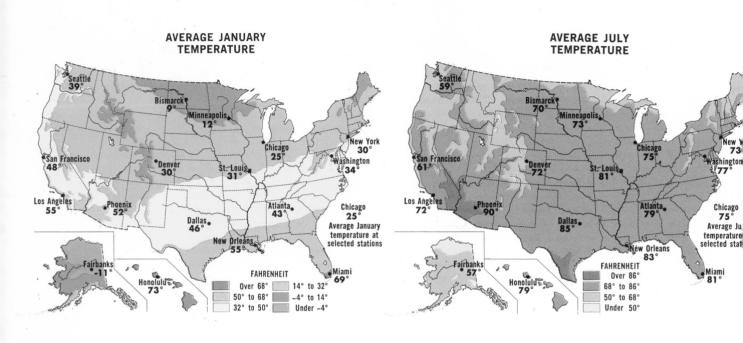

AVERAGE JANUARY TEMPERATURE

Seattle 39°
Bismarck 9°
Minneapolis 12°
Chicago 25°
New York 30°
Washington 34°
San Francisco 48°
Denver 30°
St. Louis 31°
Los Angeles 55°
Phoenix 52°
Dallas 46°
Atlanta 43°
New Orleans 55°
Miami 69°
Fairbanks -11°
Honolulu 73°

Chicago 25° Average January temperature at selected stations

FAHRENHEIT
Over 68°   14° to 32°
50° to 68°   -4° to 14°
32° to 50°   Under -4°

AVERAGE JULY TEMPERATURE

Seattle 59°
Bismarck 70°
Minneapolis 73°
Chicago 75°
New Y 73
Washington 77°
San Francisco 61°
Denver 72°
St. Louis 81°
Los Angeles 72°
Phoenix 90°
Dallas 85°
Atlanta 79°
New Orleans 83°
Miami 81°
Fairbanks 57°
Honolulu 79°

Chicago 75° Average Ju temperature selected stat

FAHRENHEIT
Over 86°
68° to 86°
50° to 68°
Under 50°

RAINFALL

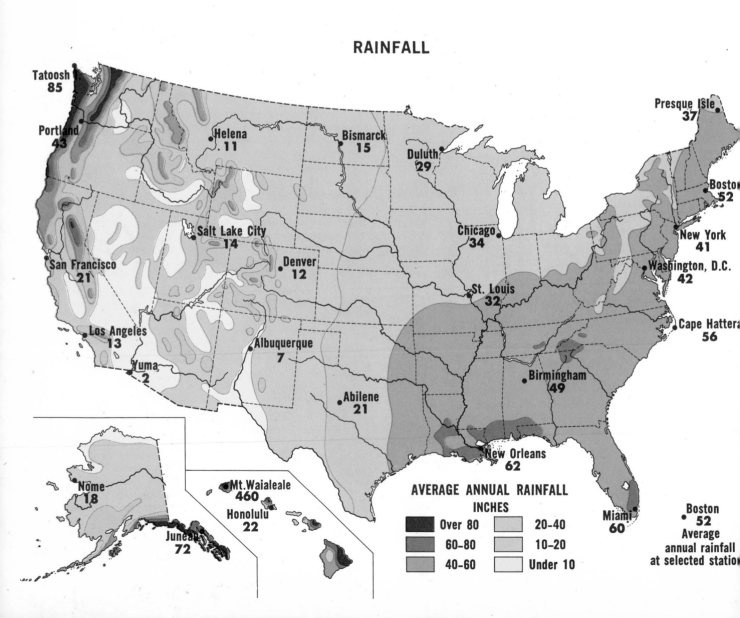

Tatoosh I. 85
Presque Isle 37
Portland 43
Helena 11
Bismarck 15
Duluth 29
Boston 52
Salt Lake City 14
Chicago 34
San Francisco 21
Denver 12
St. Louis 32
New York 41
Washington, D.C. 42
Los Angeles 13
Albuquerque 7
Cape Hattera 56
Yuma 2
Abilene 21
Birmingham 49
New Orleans 62
Nome 18
Mt. Waialeale 460
Honolulu 22
Juneau 72
Miami 60
Boston 52

AVERAGE ANNUAL RAINFALL
INCHES
Over 80   20-40
60-80   10-20
40-60   Under 10

Boston 52 Average annual rainfall at selected station

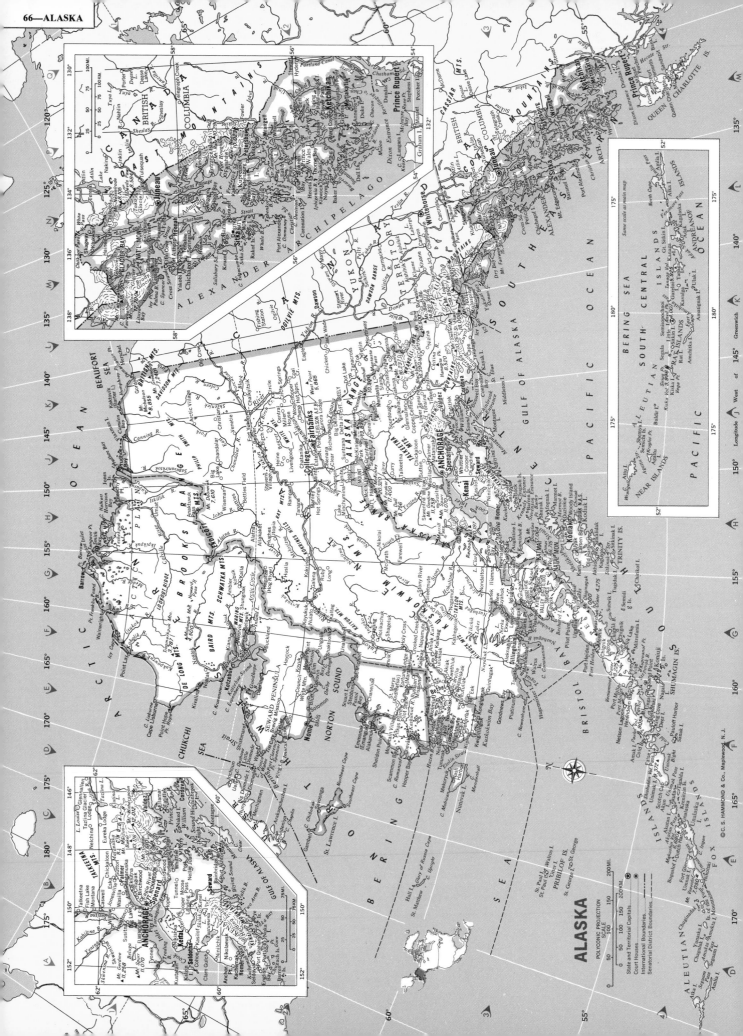

ALASKA

POLYCONIC PROJECTION
SCALE

⊛ State and Territorial Capitals.
□ Court Houses
International Boundaries.
Senatorial District Boundaries.

© C.S. HAMMOND & Co., Maplewood, N.J.

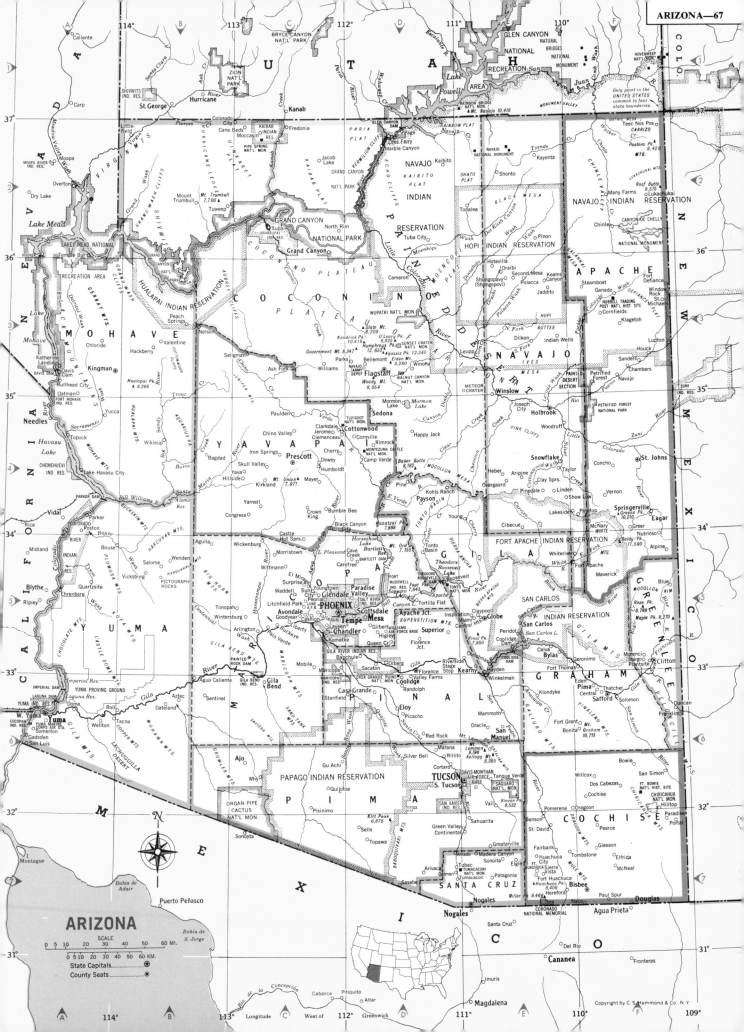

ARIZONA

SCALE
0 5 10 20 30 40 50 60 MI.
0 5 10 20 30 40 50 60 KM.

State Capitals.............⊛
County Seats.............◉

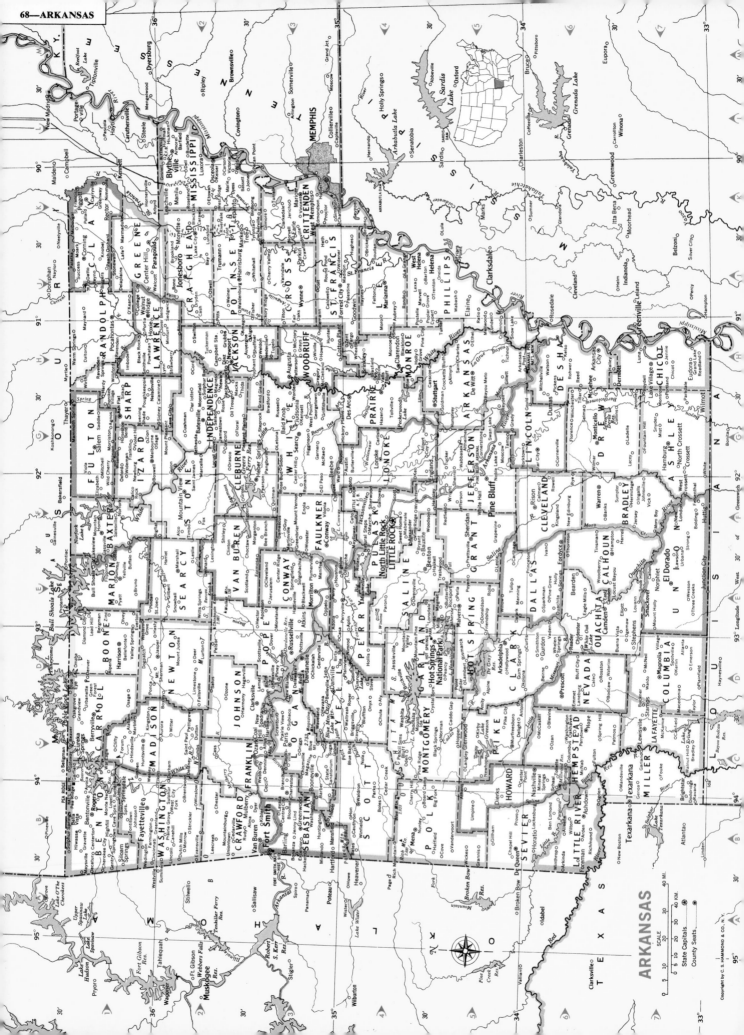

ARKANSAS

SAN FRANCISCO
AND
VICINITY

CALIFORNIA
SCALE
0 10 20   40   60   80 MI.
0 10 20 40   60   80KM.
State Capitals..........⊛
County Seats...........◉
Canals................

SACRAMENTO
AND
VICINITY

LOS ANGELES
AND VICINITY

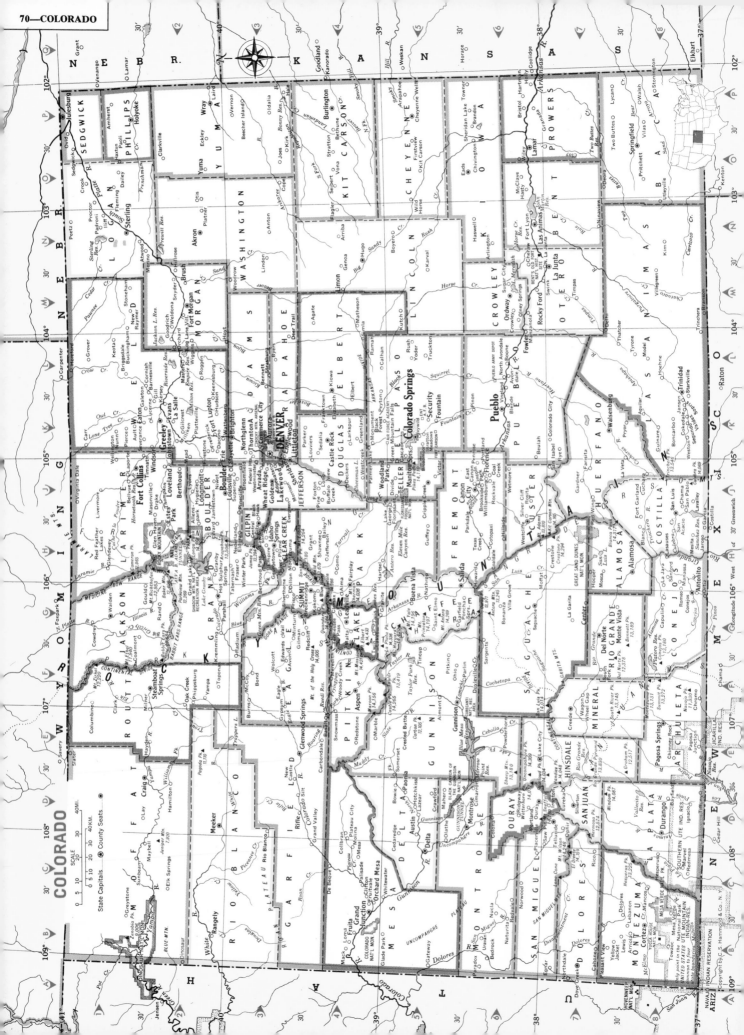

COLORADO

SCALE
40 MI.
0 10 20 30
0 5 10 20 30 40KM.

⊛ State Capitals
⊙ County Seats

CONNECTICUT

FLORIDA

SCALE
0 5 10   20   30   40   50 MI.
0 5 10 20 30 40 50 KM.

State Capitals..............⊛
County Seats..............⊙
Canals..............

FLORIDA
WESTERN PART
Same scale as main map

GEORGIA

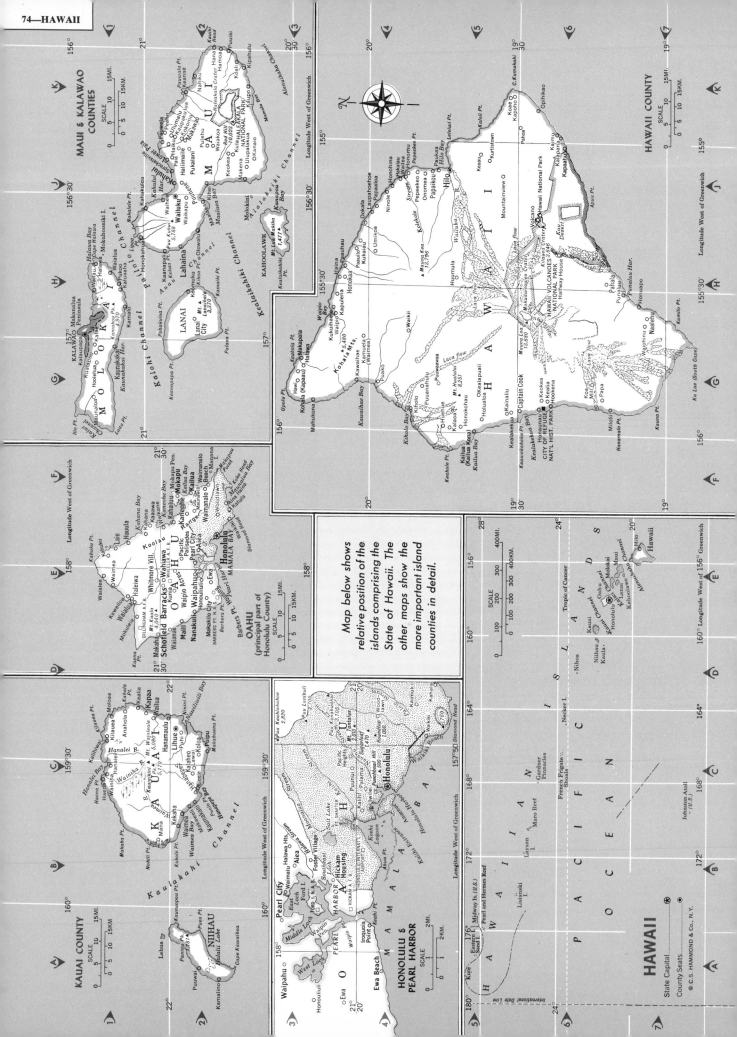

MAUI & KALAWAO COUNTIES

HAWAII COUNTY

KAUAI COUNTY

OAHU
(principal part of Honolulu County)

Map below shows relative position of the islands comprising the State of Hawaii. The other maps show the more important island counties in detail.

HONOLULU & PEARL HARBOR

HAWAII

State Capital
County Seats

© C.S. HAMMOND & Co., N.Y.

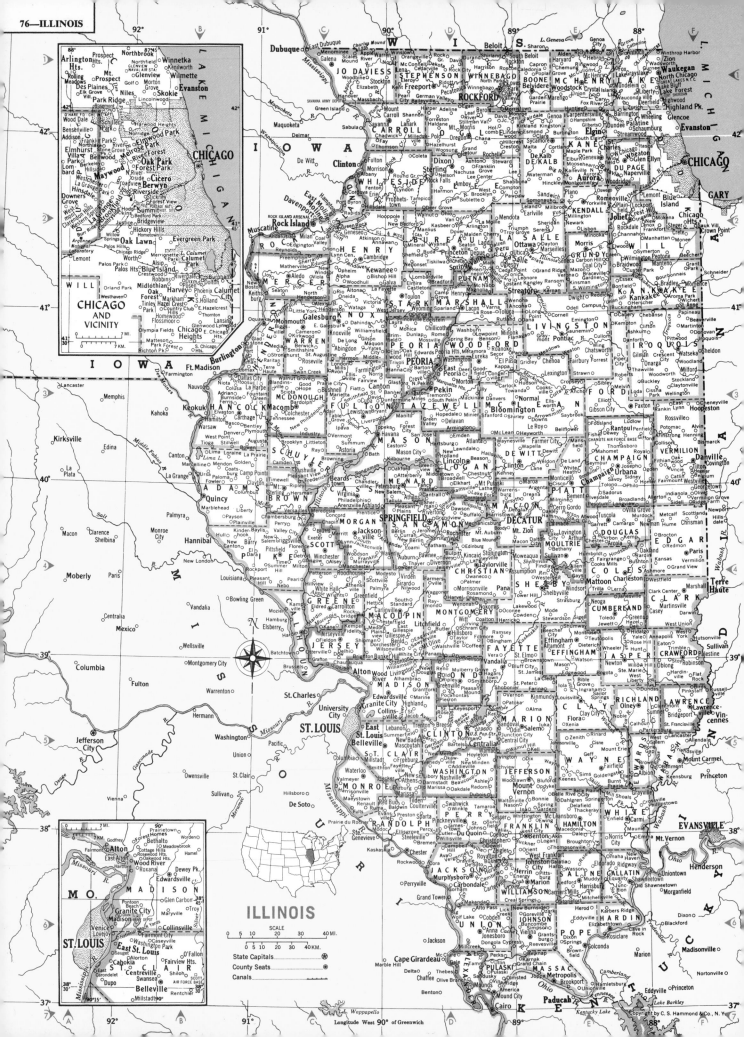

CHICAGO
AND
VICINITY

ILLINOIS

SCALE

0 5 10   20   30   40 MI.

0 10   20   30   40 KM.

State Capitals .... ⊛
County Seats .... ●
Canals

Longitude West 90° of Greenwich

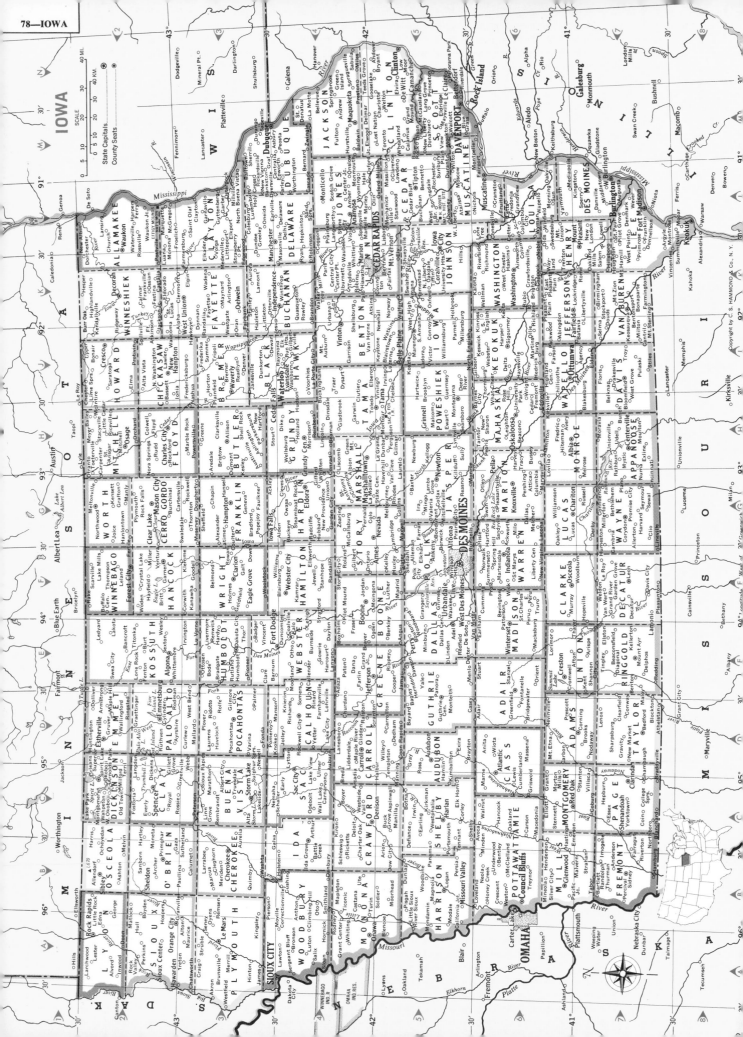

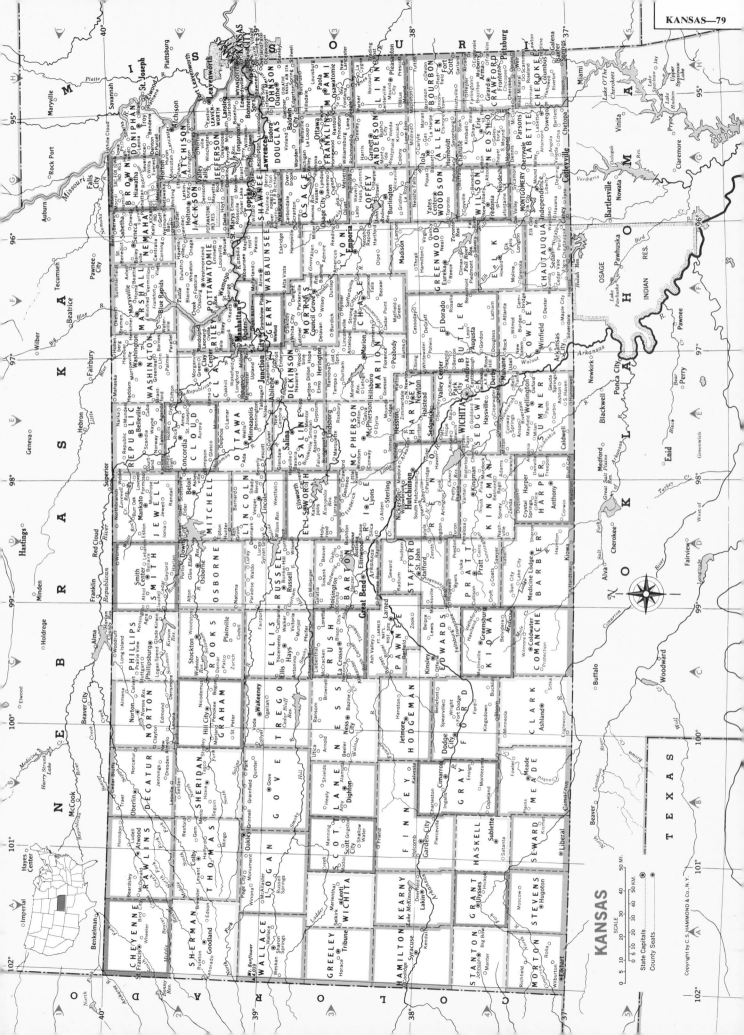

KANSAS

SCALE
State Capitals ✪
County Seats ◉

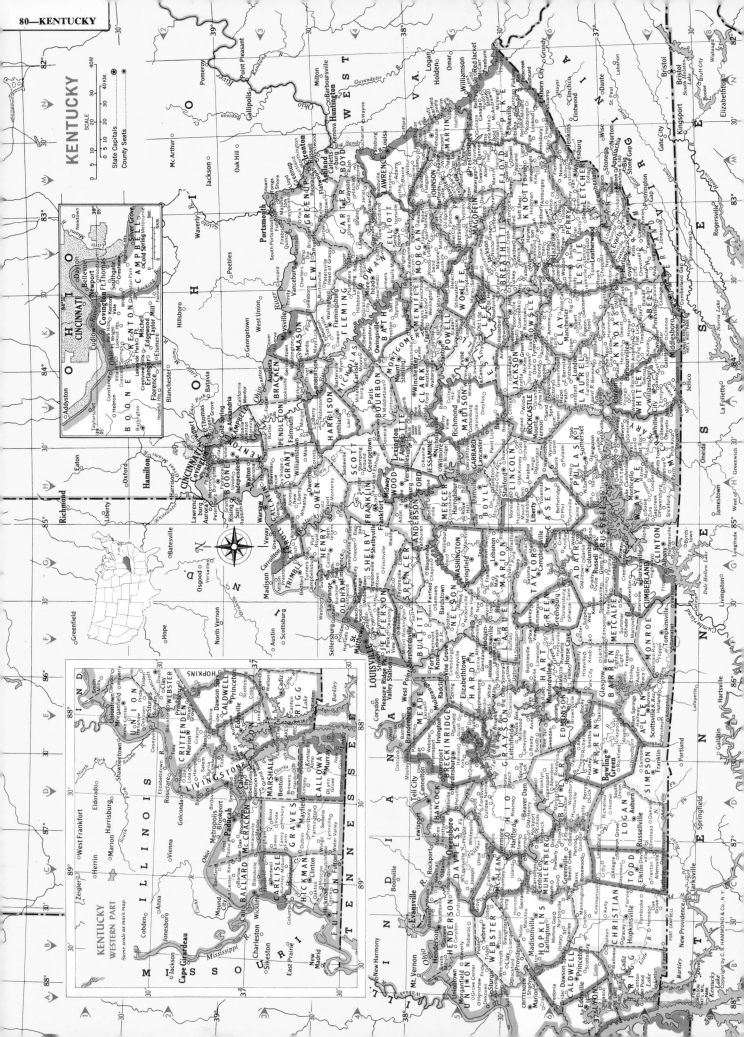

KENTUCKY

SCALE
State Capitals ⊛
County Seats ○

KENTUCKY
WESTERN PART
Same scale as main map.

Copyright by C. S. HAMMOND & Co., N.Y.

NEW ORLEANS, BATON ROUGE AND VICINITY

LOUISIANA

SCALE

State Capitals ⊛
Parish Seats ⊙
Canals ----

© C.S. HAMMOND & Co., N.Y.

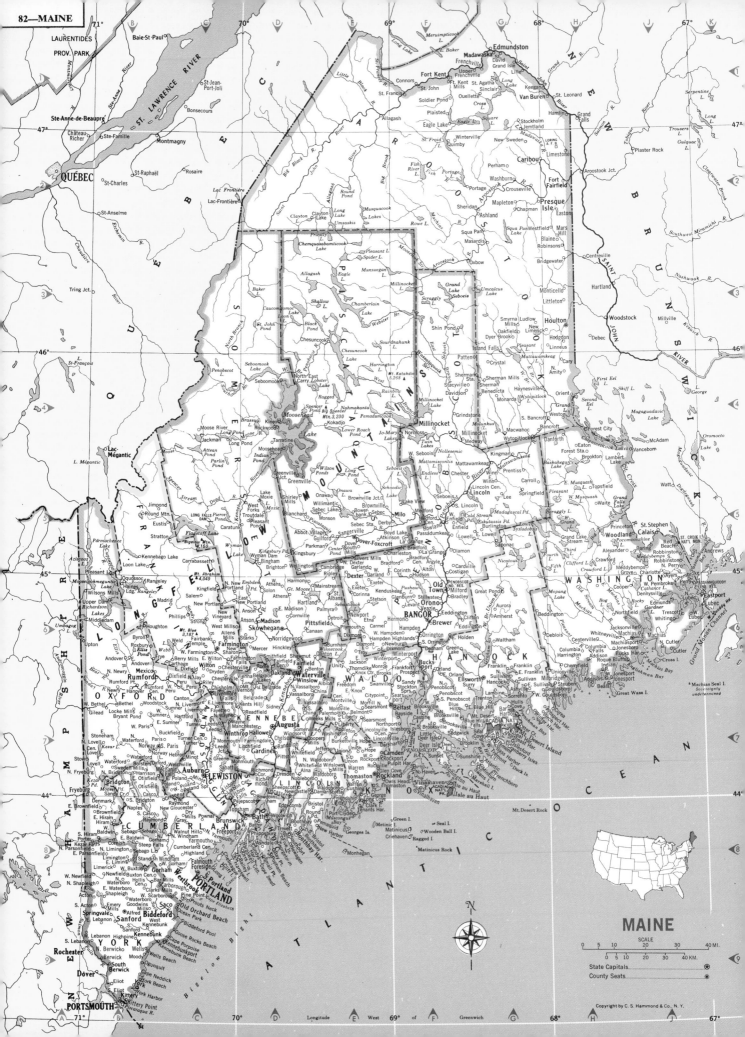

MAINE

SCALE

State Capitals .... ⊛
County Seats .... ⊛

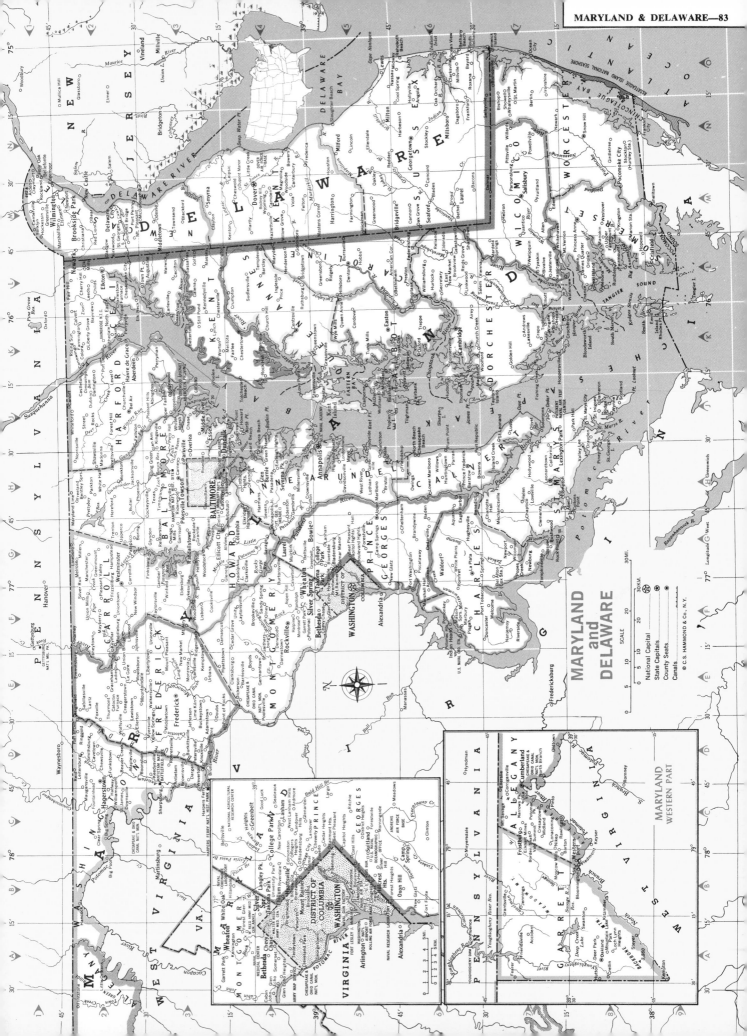

MARYLAND
and
DELAWARE

SCALE
0   5   10   20   30 MI.
0   5 10   20   30 KM.

National Capital ⊛
State Capitals ⊛
County Seats ⊙
Canals.

© C.S. HAMMOND & Co., N.Y.

MARYLAND
WESTERN PART

DISTRICT OF
COLUMBIA
WASHINGTON

MICHIGAN

SCALE
0 5 10 20 30 40 50 MI.
0 5 10 20 30 40 50 KM.

State Capitals .... ⊛
County Seats .... ⊙
Canals .... ≍

MICHIGAN WESTERN PART
as main map

LAKE SUPERIOR

LAKE MICHIGAN

LAKE HURON

LAKE ERIE

LAKE ST. CLAIR

GREEN BAY

ISLE ROYALE (NATIONAL PARK)
L. SUPERIOR
Same scale as main map

WISCONSIN

ILL.

INDIANA

OHIO

ONTARIO

CANADA

NORTH CHANNEL

DETROIT
WINDSOR

Copyright by C. S. Hammond & Co., N.Y.

Counties (selection): KEWEENAW, HOUGHTON, BARAGA, MARQUETTE, IRON, DICKINSON, MENOMINEE, DELTA, SCHOOLCRAFT, LUCE, CHIPPEWA, MACKINAC, ALGER, GOGEBIC, ONTONAGON, EMMET, CHEBOYGAN, PRESQUE ISLE, CHARLEVOIX, ANTRIM, OTSEGO, MONTMORENCY, ALPENA, LEELANAU, GRAND TRAVERSE, KALKASKA, CRAWFORD, OSCODA, ALCONA, BENZIE, MANISTEE, WEXFORD, MISSAUKEE, ROSCOMMON, OGEMAW, IOSCO, MASON, LAKE, OSCEOLA, CLARE, GLADWIN, ARENAC, HURON, OCEANA, NEWAYGO, MECOSTA, ISABELLA, MIDLAND, BAY, TUSCOLA, SANILAC, MUSKEGON, MONTCALM, GRATIOT, SAGINAW, LAPEER, ST. CLAIR, KENT, IONIA, CLINTON, SHIAWASSEE, GENESEE, OTTAWA, MACOMB, ALLEGAN, BARRY, EATON, INGHAM, LIVINGSTON, OAKLAND, WAYNE, VAN BUREN, KALAMAZOO, CALHOUN, JACKSON, WASHTENAW, CASS, ST. JOSEPH, BRANCH, HILLSDALE, LENAWEE, MONROE, BERRIEN

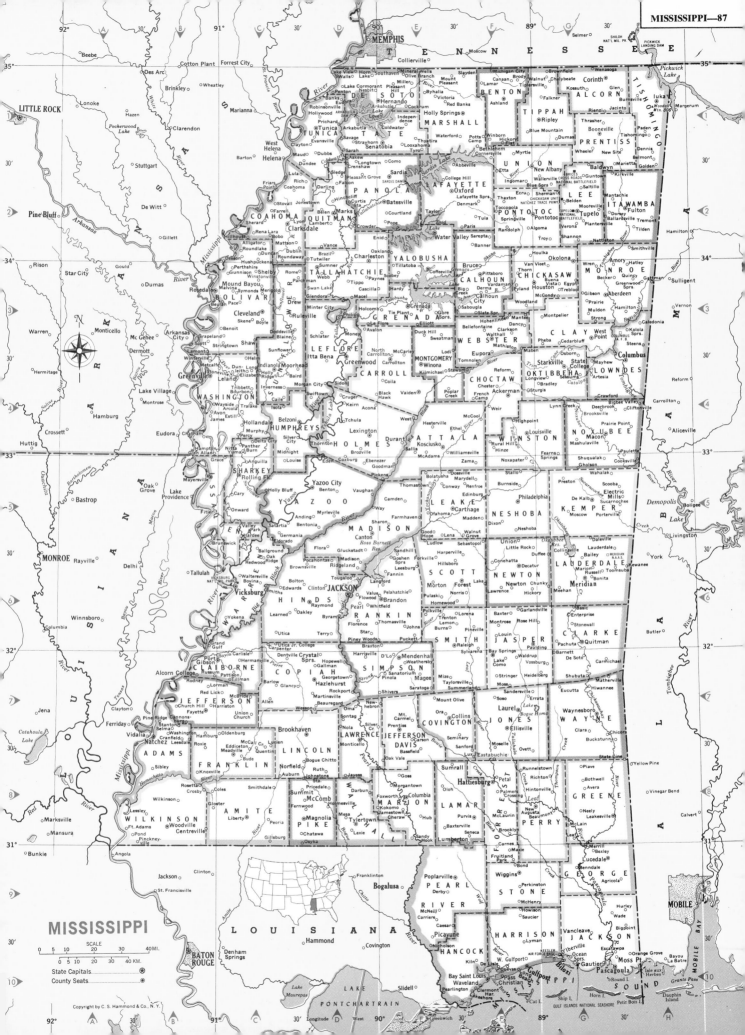

MISSISSIPPI

SCALE
0 5 10   20   30   40 MI.
0 5 10   20   30   40 KM.

State Capitals............⊛
County Seats.............○

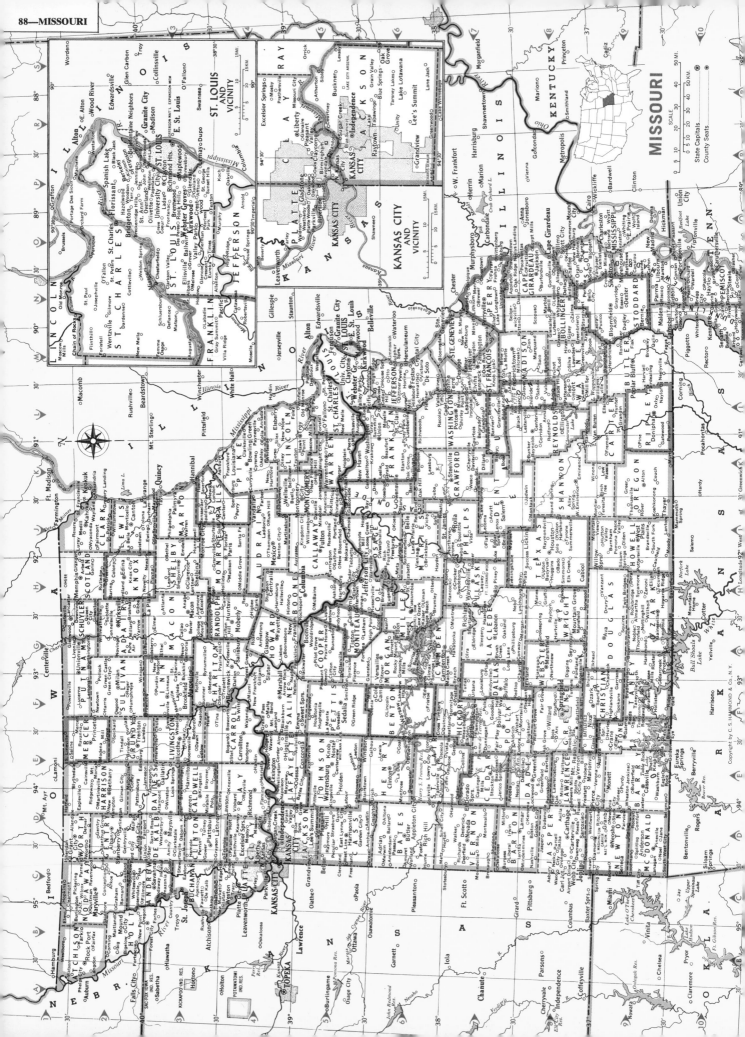

MISSOURI

State Capitals ◎
County Seats ⊙

ST. LOUIS
AND
VICINITY

KANSAS CITY
AND
VICINITY

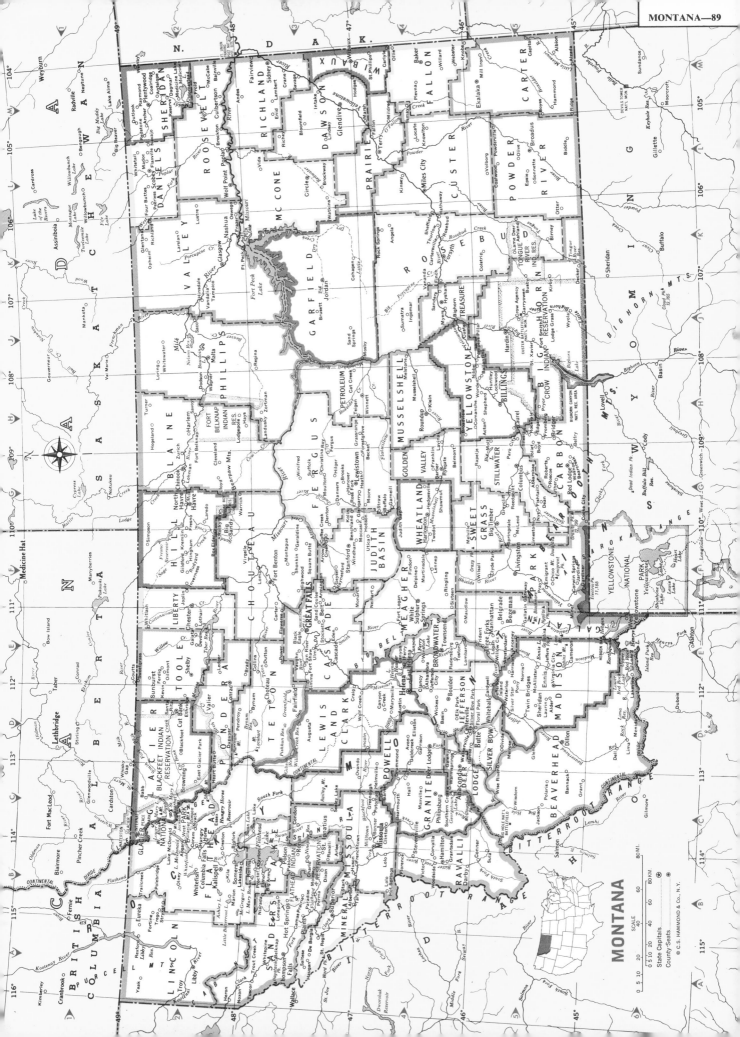

MONTANA

SCALE
0 10 20   40   60   80 MI.
0 5 10 20   40   60 80 KM

⊛ State Capitals
⊛ County Seats

© C.S. HAMMOND & CO., N.Y.

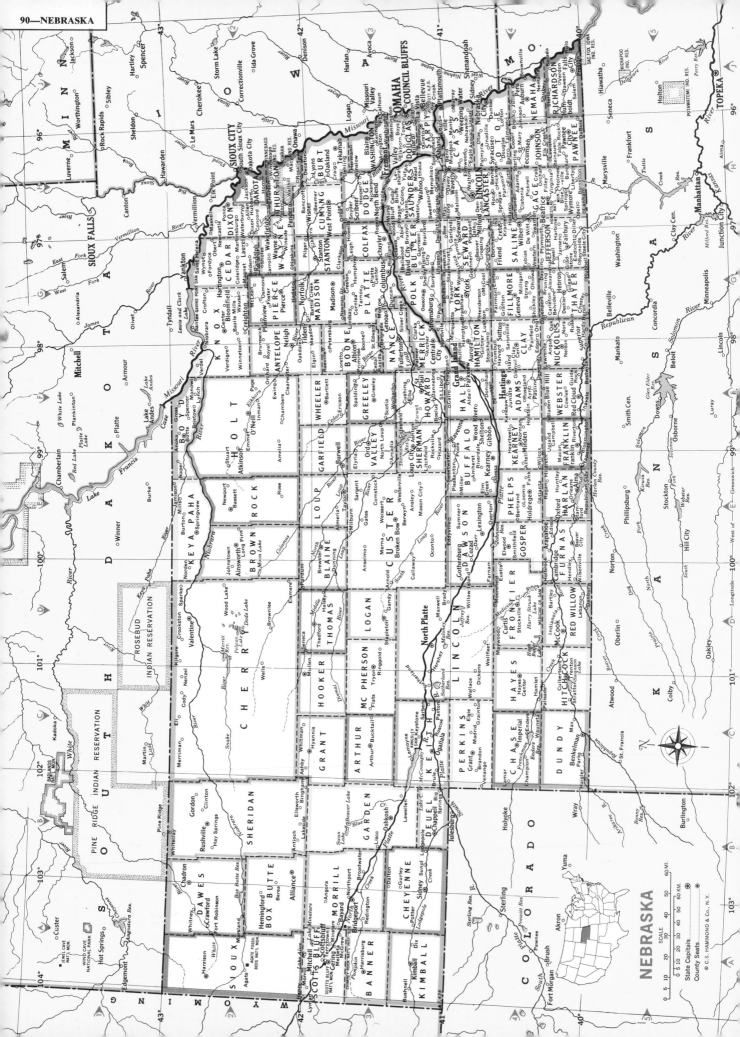

NEBRASKA

SCALE
0 10 20 30 40 50 60 MI.
0 5 10 20 30 40 50 60 KM.
◉ State Capitals
⊛ County Seats

© C.S. HAMMOND & Co., N.Y.

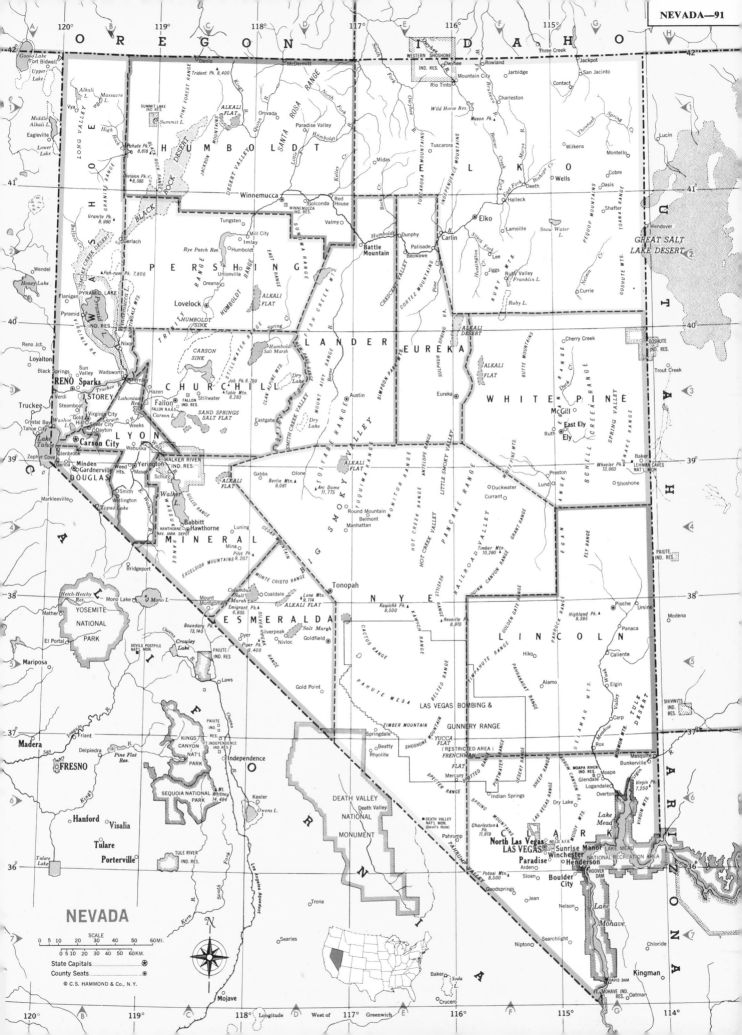

NEVADA

SCALE
0 5 10   20   30   40   50   60 MI.
0 5 10   20   30   40   50   60 KM.

State Capitals.................................⊛
County Seats..................................◉

© C.S. HAMMOND & Co., N.Y.

NEW HAMPSHIRE

SCALE
0 5 10 15 20 25 MI.
0 5 10 15 20 25 KM.

State Capitals ....⊛
County Seats ....◉

NEW JERSEY

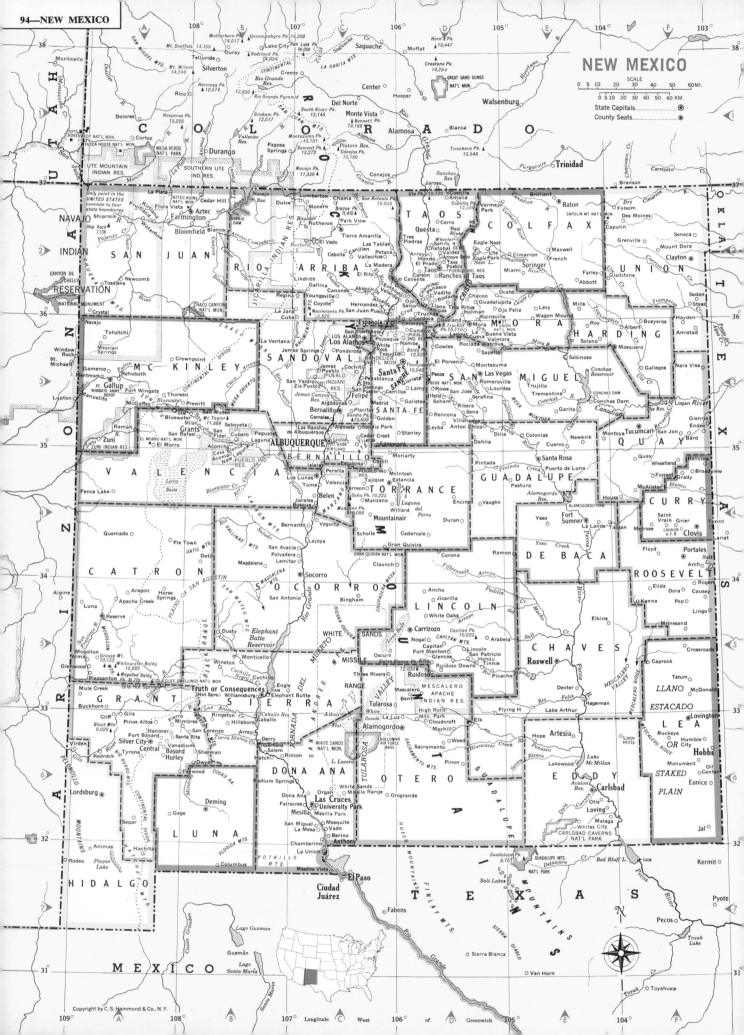

NEW MEXICO

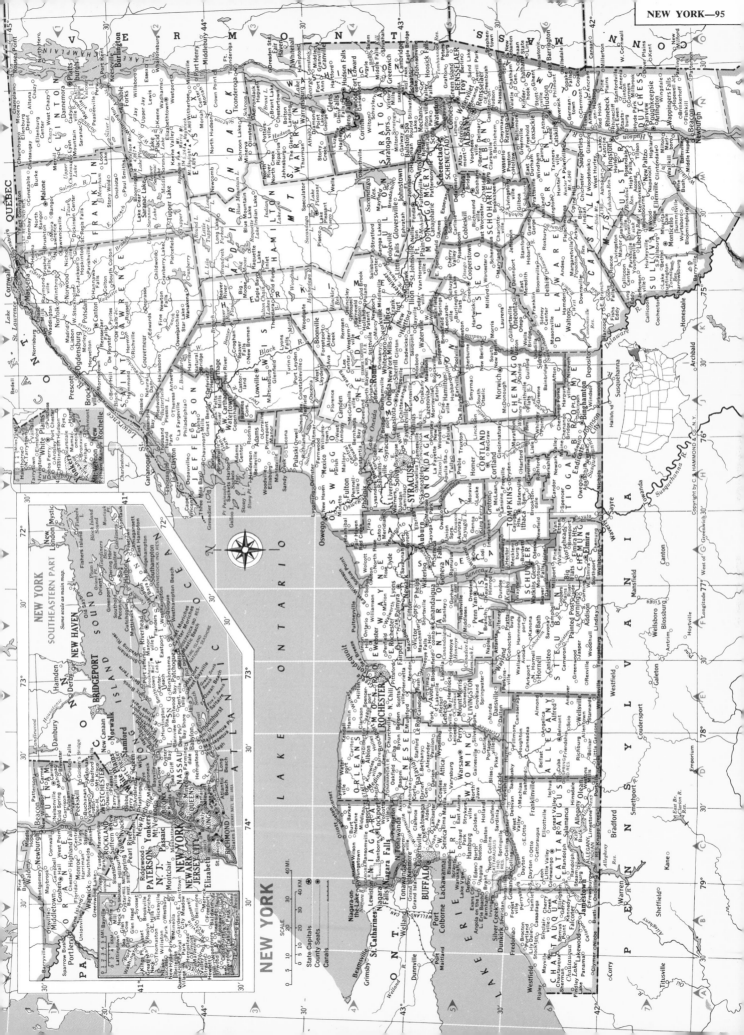

NEW YORK

SCALE
40 MI.
40 KM

State Capitals
County Seats
Canals

NEW YORK
SOUTHEASTERN PART
Some scale as main map.

NORTH
CAROLINA

SCALE
State Capitals
County Seats
Canals

NORTH CAROLINA
WESTERN PART
Same scale as main map.

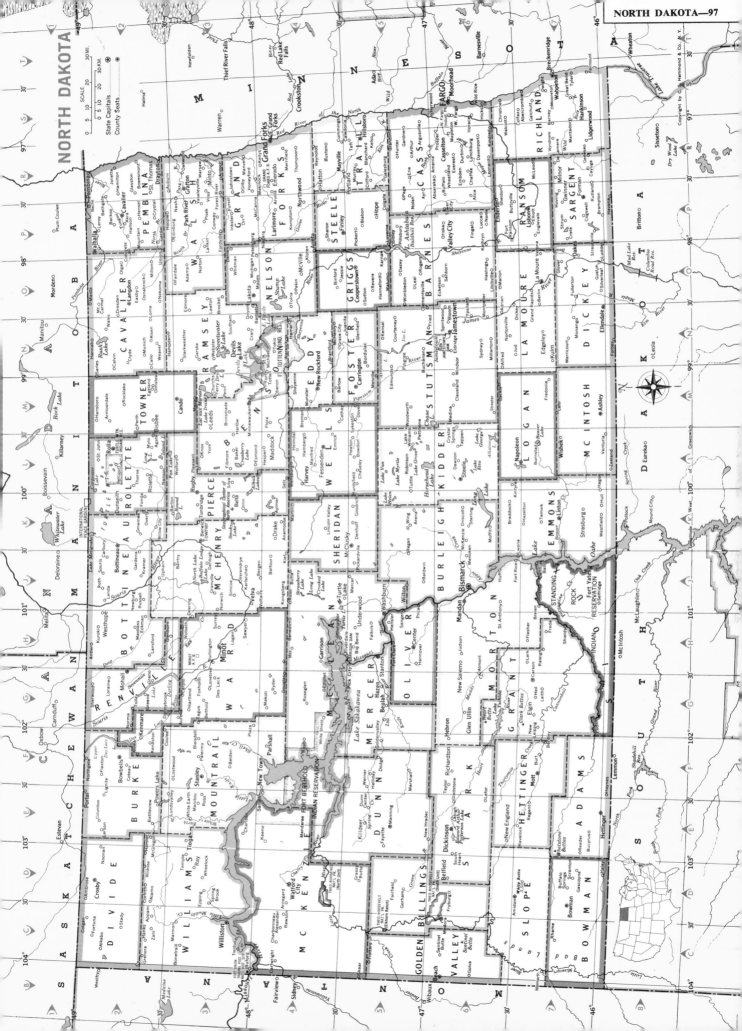

OHIO

SCALE
0 5 10   20   30   40 MI.
0 5 10 20   30   40 KM.

State Capitals .... ●
County Seats .... ○

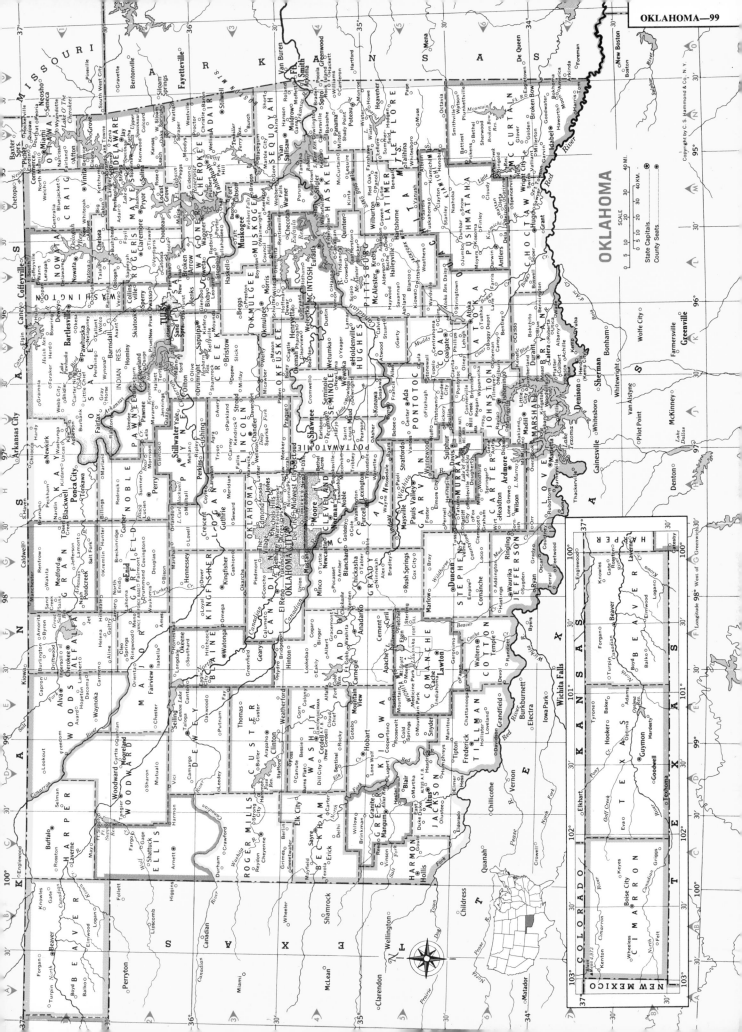

OKLAHOMA

OREGON

SCALE

0 5 10 20 30 40 50 60 MI.
0 5 10 20 30 40 50 60 KM.

⊛ State Capitals
◉ County Seats

Copyright by C. S. Hammond & Co., N. Y.

PORTLAND, SALEM
AND
VICINITY

PENNSYLVANIA

SCALE
0   5   10   20   30   40MI.
0   5 10   20   30   40KM.

⊛ State Capitals.
⊙ County Seats.
     Canals.

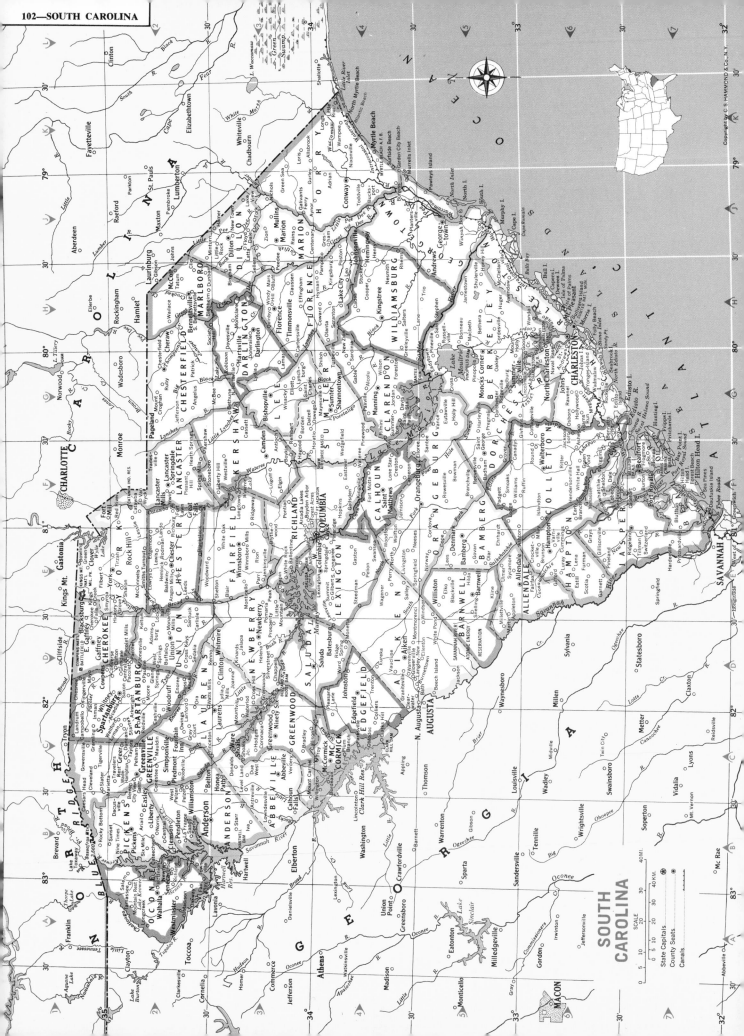

SOUTH
CAROLINA

SCALE
0 5 10   20   30   40 MI.
0 5 10 20   30   40 KM.

State Capitals........... ⊛
County Seats............. ◉
Canals..................

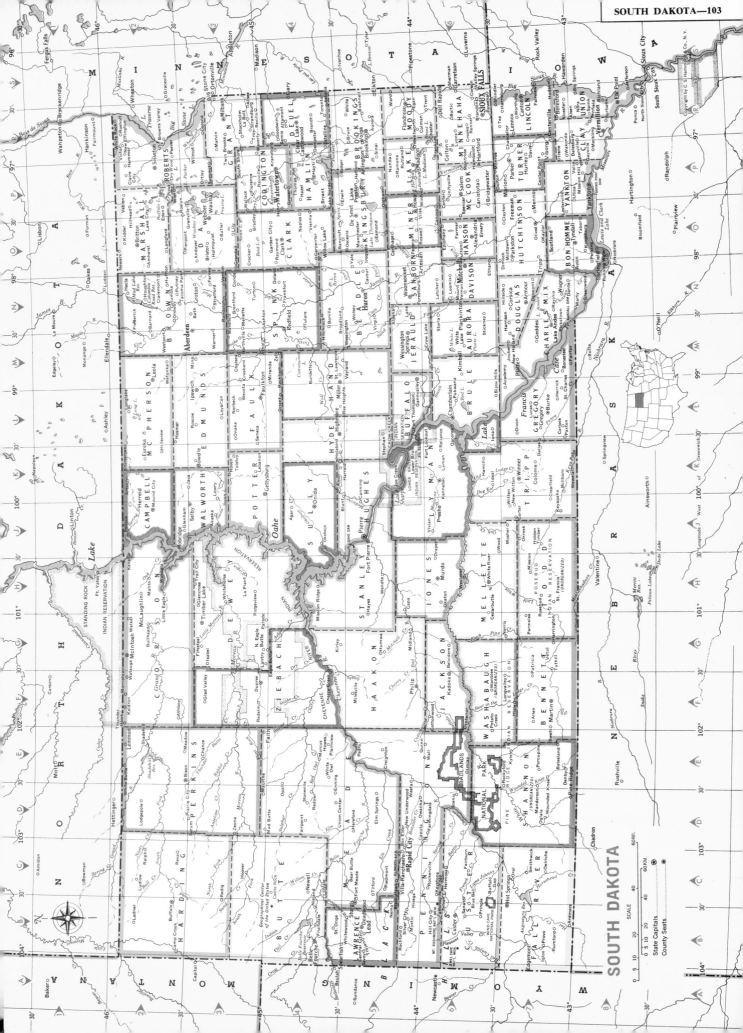

SOUTH DAKOTA

SCALE
0   10   20   30   40   60 MI.
0 5 10   20   30   40   60 KM.

State Capitals ⊛
County Seats ◉

TENNESSEE

SCALE

State Capitals
County Seats

S—95111—

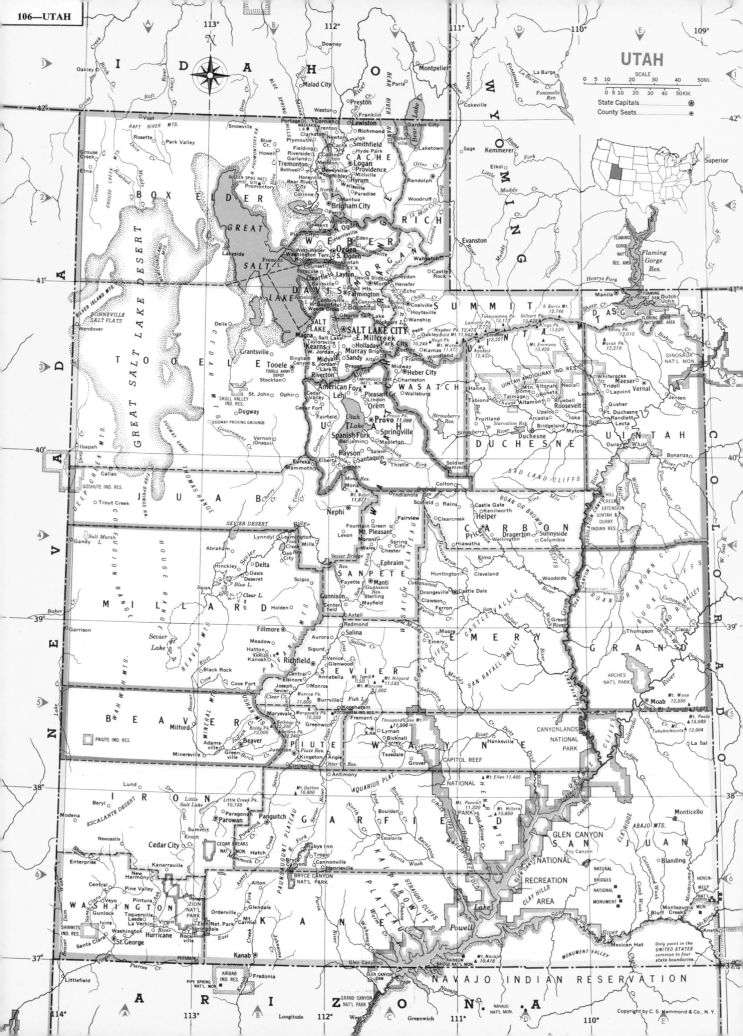

UTAH

SCALE

0 5 10   20   30   40   50 MI.

0 5 10   20   30   40   50 KM.

State Capitals .... ⊛
County Seats .... ◉

Copyright by C. S. Hammond & Co., N.Y.

Only point in the
UNITED STATES
common to four
state boundaries.

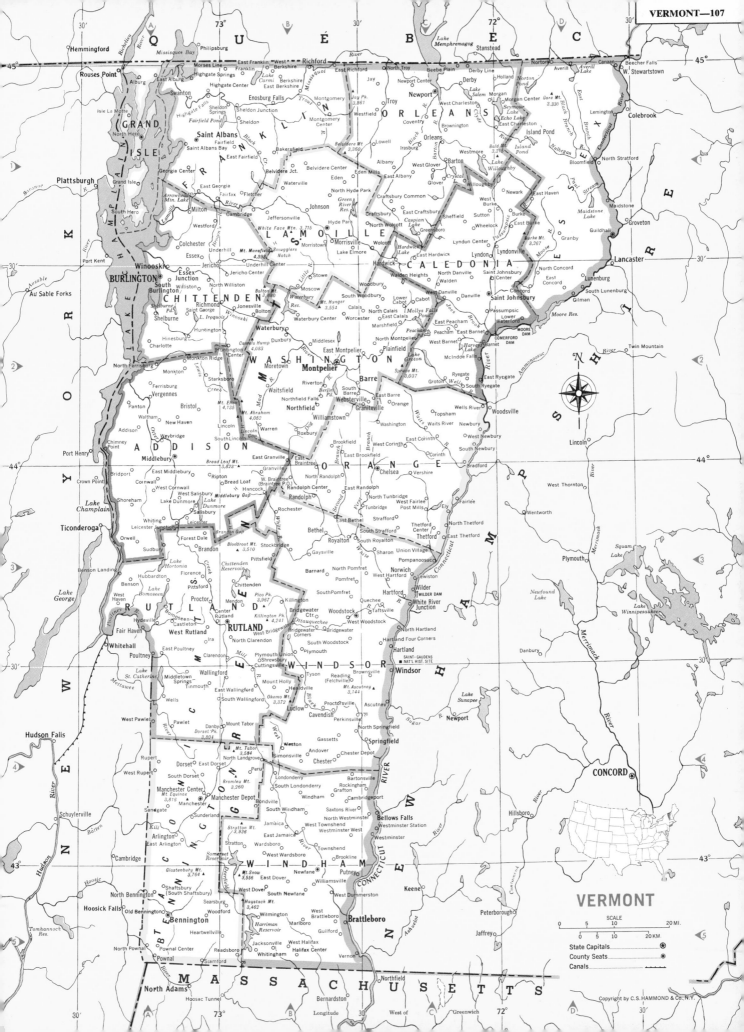

VERMONT

SCALE
0   5   10   20 MI.
0 5 10   20 KM.

State Capitals............................ ⊛
County Seats............................. ⊚
Canals

Copyright by C.S. HAMMOND & Co. N.Y.

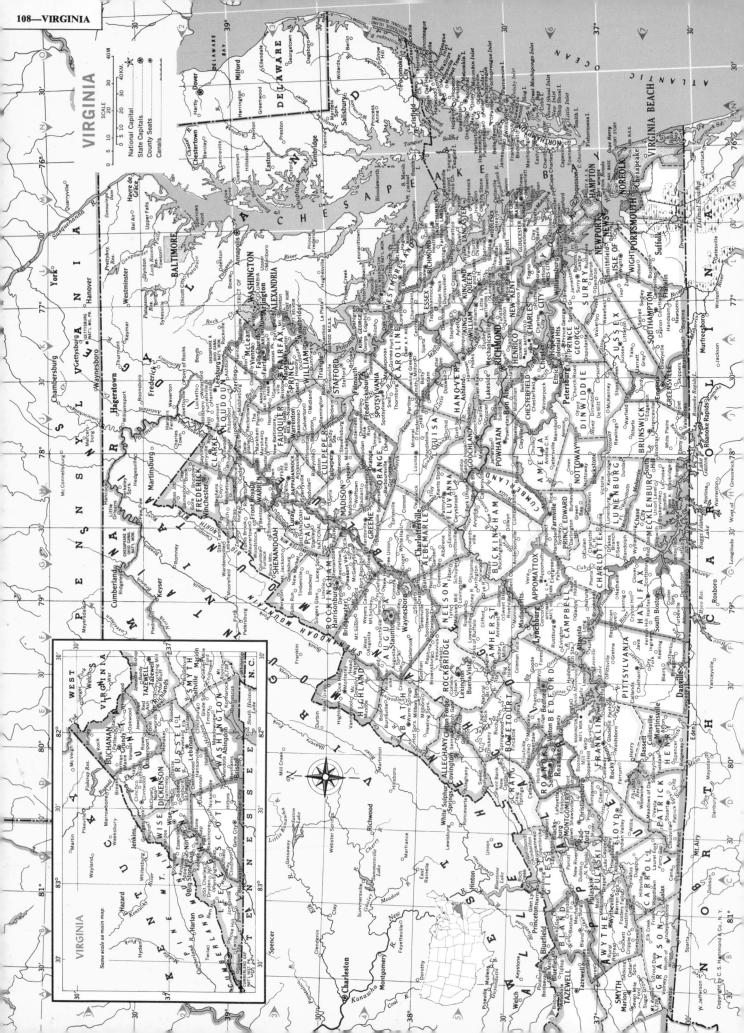

VIRGINIA

SCALE

National Capital ⊛
State Capitals ⊛
County Seats ⊙
Canals

WASHINGTON

SCALE
0 5 10 20 30 40 MI.
0 5 10 20 30 40KM.

State Capitals ⊛
County Seats ⊙

Copyright by C.S. HAMMOND & Co., N.Y.

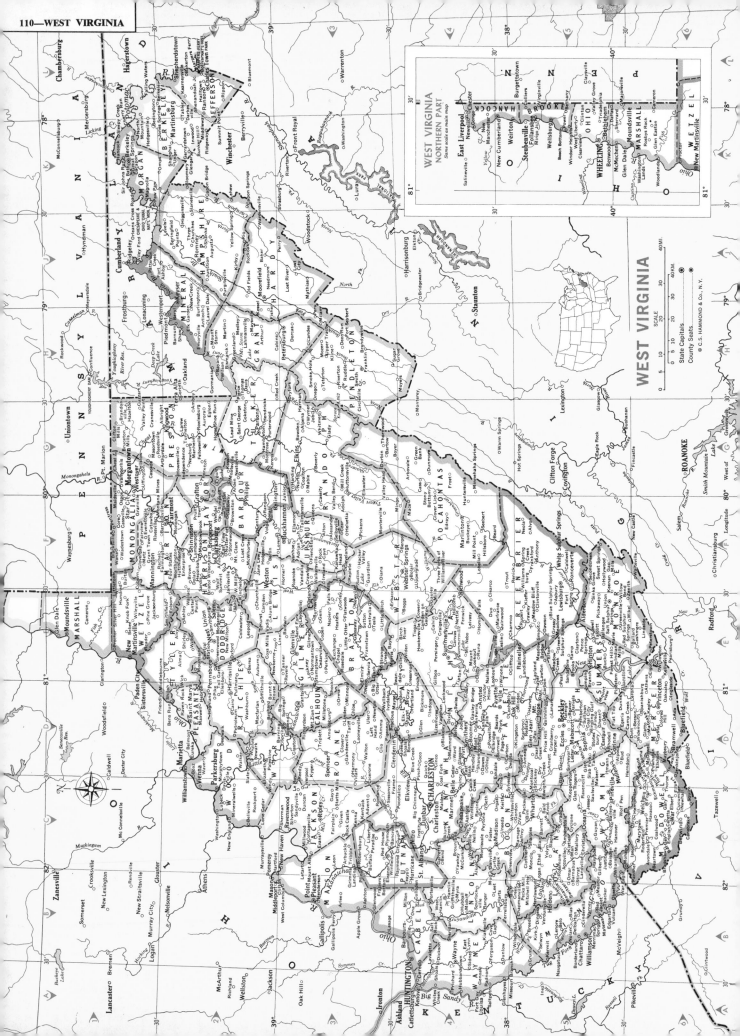

WEST VIRGINIA

WEST VIRGINIA
NORTHERN PART
Same scale as main map

SCALE

State Capitals
County Seats

© C.S. HAMMOND & CO., N.Y.

WISCONSIN

LAKE SUPERIOR

APOSTLE ISLANDS
APOSTLE ISLANDS NAT'L LAKESHORE

Duluth
Superior
St. Paul

LAKE MICHIGAN

MILWAUKEE

MADISON

Green Bay
Oshkosh
Fond du Lac
Appleton
Sheboygan
Manitowoc
Racine
Kenosha
Eau Claire
La Crosse
Wausau
Stevens Point
Wisconsin Rapids
Rhinelander
Eagle River
Sturgeon Bay

Counties: DOUGLAS, BAYFIELD, ASHLAND, IRON, VILAS, FLORENCE, BURNETT, WASHBURN, SAWYER, PRICE, ONEIDA, FOREST, MARINETTE, POLK, BARRON, RUSK, LINCOLN, LANGLADE, OCONTO, ST. CROIX, DUNN, CHIPPEWA, TAYLOR, MARATHON, MENOMINEE, SHAWANO, PIERCE, PEPIN, EAU CLAIRE, CLARK, WOOD, PORTAGE, WAUPACA, OUTAGAMIE, BROWN, KEWAUNEE, BUFFALO, TREMPEALEAU, JACKSON, WAUSHARA, WINNEBAGO, CALUMET, MANITOWOC, LA CROSSE, MONROE, JUNEAU, ADAMS, MARQUETTE, GREEN LAKE, FOND DU LAC, SHEBOYGAN, VERNON, RICHLAND, SAUK, COLUMBIA, DODGE, WASHINGTON, OZAUKEE, CRAWFORD, GRANT, IOWA, DANE, JEFFERSON, WAUKESHA, MILWAUKEE, LAFAYETTE, GREEN, ROCK, WALWORTH, RACINE, KENOSHA

MICHIGAN
ILLINOIS
IOWA
MINNESOTA

CHICAGO

SCALE
0 5 10 20 30 40 MI.
0 5 10 20 30 40 KM.

State Capitals ⊛
County Seats ⊙
Canals

Inset map (Milwaukee area): WAUKESHA, MILWAUKEE, WASHINGTON, OZAUKEE, WALWORTH, RACINE, KENOSHA, JEFFERSON counties; Menomonee Falls, Brookfield, Wauwatosa, West Allis, Milwaukee, West Milwaukee, Cudahy, South Milwaukee, Oak Creek, Franklin, New Berlin, Greenfield, Hales Corners, Greendale, Saint Francis, General Mitchell Field, Whitefish Bay, Shorewood, Glendale, Brown Deer, Bayside, Fox Point, River Hills

Copyright by C. S. Hammond & Co., N.Y.

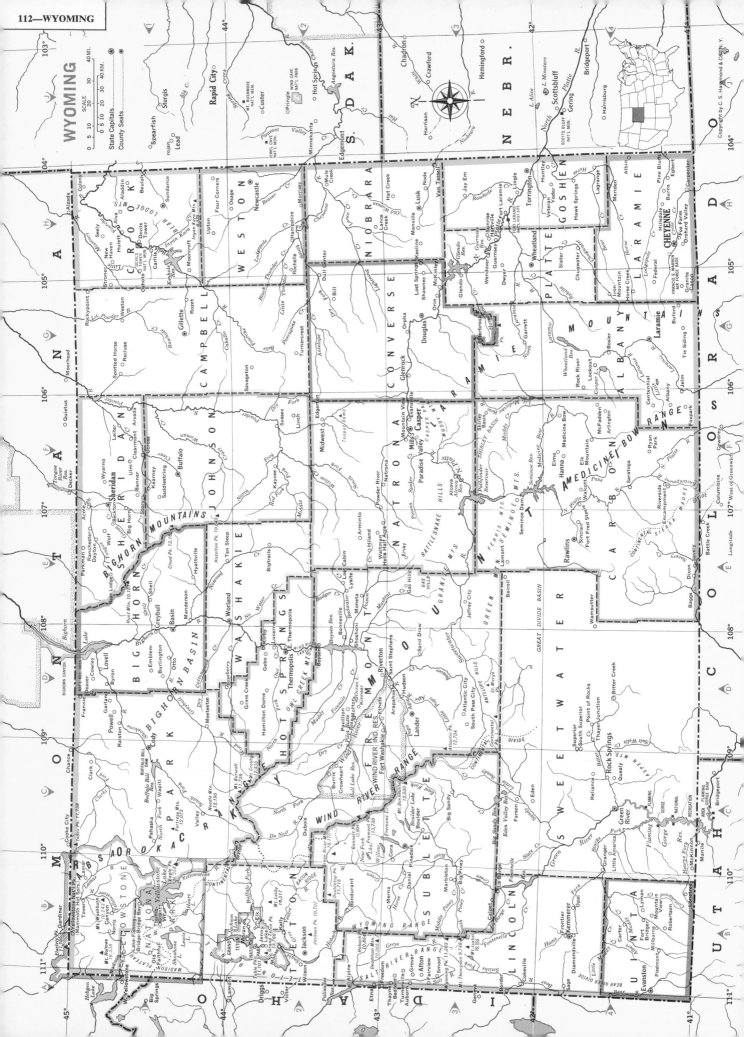

INDEX OF THE WORLD

(See Glossary of Abbreviations on Page 145)

Name	Index Ref.	Plate No.
Apurímac (riv.), Peru	F 6	42
'Aqaba (gulf), Asia	C 4	33
'Aqaba, Jordan	D 6	31
Arabia (pen.), Asia	H 7	32
Arabian (sea), Asia	K 8	32
Arabian (des.), Egypt	N 6	38
Aracaju, Braz.	N 6	42
Arad, Rum.	C 2	29
Arafura (sea)	E 1	40
Aragón (reg.), Spain	F 2	27
Araks (riv.), Asia	E 2	33
Aral (sea), U.S.S.R.	F 5	30
Aran (isls.), Ire.	B 4	23
Aransas Pass, Tex.	G10	105
Ararat (mt.), Turkey	D 1	33
Arcachon, France	C 5	26
Arcadia, Calif.	C10	69
Arcadia, Fla.	E 4	72
Arcata, Calif.	A 3	69
Archangel, U.S.S.R.	E 3	30
Archbald, Pa.	M 2	101
Arches N.P., Utah	E 5	106
Arctic Circle	17, 20	
Arctic Ocean	20	
Arden-Arcade, Calif.	B 8	69
Ardennes (plat.), Belg.	J 5	22
Ardmore, Okla.	H 6	99
Arecibo, P.R.	E 3	45
Arezzo, Italy	C 3	28
Argenteuil, France	A 1	26
Argentina		43
Arges (riv.), Rum.	D 2	29
Argonne N. Lab., Ill.	A 2	76
Árgos, Greece	C 4	29
Århus, Den.	B 3	24
Arica, Chile	F 7	42
Arizona (state), U.S.		67
Arkadelphia, Ark.	D 5	68
Arkansas (riv.), U.S.	H 3	61
Arkansas (state), U.S.		68
Arkansas City, Kans.	E 4	79
Arkansas Post N. Mem., Ark.	H 5	68
Arles, France	F 6	26
Arlington, Mass.	C 6	84
Arlington, Tex.	F 2	105
Arlington, Va.	K 3	108
Arlington Hts., Ill.	A 1	76
Armagh, N.Ire.	C 3	23
Armavir, U.S.S.R.	E 5	30
Armenian S.S.R., U.S.S.R.	E 6	30
Armentières, France	E 2	26
Armidale, Aust.	J 6	40
Arnhem, Neth.	G 4	24
Arno (riv.), Italy	C 3	28
Arnold, Pa.	C 4	101
Arran (isl.), Scot.	D 3	23
Arras, France	E 2	26
Arroyo Grande, Calif.	E 8	69
Artesia, Calif.	C11	69
Artesia, N. Mex.	E 6	94
Arthabaska, Que.	F 3	53
Artigas, Urug.	J10	43
Aruba (isl.), Neth. Ant.	E 4	45
Arusha, Tanz.	O12	39
Arvada, Colo.	J 3	70
Arvida, Que.	F 1	53
Arvika, Sweden	C 3	24
Arvin, Calif.	G 8	69
Asbury Park, N.J.	F 3	93
Ascension (isl.), St. Helena	E15	39
Ashanti (reg.), Ghana	F10	38
Ashdod, Isr.	B 4	31
Asheboro, N.C.	F 3	96
Asheville, N.C.	E 8	96
Ashkhabad, U.S.S.R.	F 6	30
Ashland, Ky.	M 4	80
Ashland, Mass.	J 3	84
Ashland, Ohio	F 4	98
Ashland, Oreg.	E 5	100
Ashland, Wis.	E 2	111
Ashqelon, Isr.	A 4	31
Ashtabula, Ohio	J 2	98
Asia		32
'Asir (reg.), Saudi Ar.	D 6	33
Asmara, Eth.	O 9	38
Aspen, Colo.	F 4	70
Assam (state), India	G 3	34
Assateague Isl. N. Sea., U.S.	O 8	83
Assiniboine (riv.), Can.	A 4	56
Asti, Italy	B 2	28
Astoria, Oreg.	D 1	100
Astrakhan', U.S.S.R.	E 5	30
Asturias (reg.), Spain	C 1	27
Asunción (cap.), Par.	J 9	43
Aswân, Egypt	N 7	38
Atacama (des.), Chile	G 8	43
Atchison, Kans.	G 2	79
Athabasca (lake), Can.	F 5	47
Athens, Ala.	E 1	65
Athens, Ga.	F 3	73
Athens (cap.), Greece	D 4	29
Athens, Ohio	F 7	98
Athens, Tenn.	M 4	104
Athens, Tex.	J 5	105
Atherton, Calif.	K 3	69
Athol, Mass.	F 2	84
Áthos (mt.), Greece	D 3	29
Atitlán (lake), Guat.	B 2	46
Atlanta (cap.), Ga.	D 3	73
Atlanta, Tex.	K 4	105
Atlantic, Iowa	D 6	78
Atlantic City, N.J.	E 5	93
Atlantic Highlands, N.J.	F 3	93
Atlantic Ocean	H-J 3-8	19
Atlas (mts.), Afr.	E 5	38
Atmore, Ala.	C 8	65
Attalla, Ala.	F 2	65
Attica, N.Y.	D 5	95
Attleboro, Mass.	J 5	84
Attu (isl.), Alaska	H 3	66
Atwater, Calif.	E 6	69
Auburn, Ala.	H 5	65
Auburn, Calif.	C 8	69
Auburn, Ind.	G 2	77
Auburn, Maine	C 7	82
Auburn, Mass.	G 4	84
Auburn, N.Y.	G 5	95
Auburn, Wash.	C 3	109
Auburndale, Fla.	E 3	72
Auckland, N.Z.	L 5	40
Audubon, N.J.	B 3	93
Augsburg, W.Ger.	D 4	25
Augusta, Ga.	J 4	73
Augusta, Kans.	F 4	79
Augusta (cap.), Maine	D 7	82
Aurora, Colo.	K 3	70
Aurora, Ill.	E 2	76
Aurora, Mo.	E 9	88
Aurora, Ohio	H 3	98
Austin, Minn.	E 7	86
Austin (cap.), Tex.	G 7	105
Austintown, Ohio	J 3	98
Austral (isls.), Fr. Poly.	L 8	41
Australia		40
Australian Alps (mts.), Aust.	H 7	40
Austria		29
Auvergne (mts.), France	E 5	26
Avalon (pen.), Newf.	D 2	50
Avalon, Pa.	B 6	101
Avellaneda, Arg.	O12	43
Avignon, France	F 6	26
Ávila, Spain	D 2	27
Avon, Conn.	D 1	71
Avon (riv.), Eng.	F 4	23
Avon, Mass.	K 4	84
Avon, Ohio	F 3	98
Avondale, Ariz.	C 5	67
Avon Lake, Ohio	F 2	98
Avon Park, Fla.	E 4	72
Ayacucho, Peru	F 6	42
Ayer, Mass.	H 2	84
Ayers Rock (mt.), Aust.	E 5	40
Ayr, Scot.	D 3	23
Ayutthaya, Thai.	D 4	35
Azcapotzalco, Mex.	F 1	46
Azerbaidzhan S.S.R., U.S.S.R.	E 5	30
Azores (isls.), Port.	H 4	19
Azov (sea), U.S.S.R.	D 5	30
Aztec Ruins N.M., N. Mex.	A 2	94
Azusa, Calif.	D10	69

B

Name	Index Ref.	Plate No.
Bab el Mandeb (str.)	D 7	33
Babylon, N.Y.	E 2	95
Bacău, Rum.	D 2	29
Bacolod, Phil.	G 3	36
Badajoz, Spain	C 3	27
Baden, Pa.	B 4	101
Baden-Baden, W.Ger.	B 4	25
Baden-Württemberg (state), W.Ger.	C 4	25
Badgastein, Aus.	B 3	29
Bad Kissingen, W.Ger.	D 3	25
Badlands N.P., S. Dak.	E 6	103
Baffin (isl.), N.W.T.	L 2	60
Baghdad (cap.), Iraq	D 3	33
Bahamas	C 1	45
Bahia (Salvador), Braz.	N 6	42
Bahía Blanca, Arg.	G11	43
Bahrain	F 4	33
Bainbridge, Ga.	C 9	73
Bairiki (cap.), Kiribati	H 5	41
Baja California (states), Mex.	A-B 1-3	46
Baker, La.	K 1	81
Baker, Oreg.	K 3	100
Baker (mt.), Wash.	D 2	109
Bakersfield, Calif.	G 8	69
Baku, U.S.S.R.	F 5	30
Balaton (lake), Hung.	D 3	29
Balch Sprs., Tex.	H 2	105
Baldwin, N.Y.	B 3	95
Baldwin, Pa.	B 7	101
Baldwin Park, Calif.	D10	69
Baldwinsville, N.Y.	H 4	95
Balearic (isls.), Spain	G 3	27
Bali (isl.), Indon.	E 7	36
Balkan (mts.), Bulg.	C 3	29
Balkh, Afgh.	B 1	34
Balkhash (lake), U.S.S.R.	H 5	30
Ballarat, Aust.	G 7	40
Ballina, Ire.	B 3	23
Ballwin, Mo.	O 3	88
Balmoral Castle, Scot.	E 2	23
Baltic (sea), Europe	L 4	22
Baltimore, Ire.	B 5	23
Baltimore, Md.	H 3	83
Baltistan (reg.), India	D 1	34
Baluchistan (reg.), Pak.	A 3	34
Bamako (cap.), Mali	E 9	38
Bamberg, W.Ger.	D 4	25
Banaba (isl.), Kiribati	G 6	41
Banda (sea), Indon.	H 6	36
Bandar 'Abbas, Iran	G 4	33
Bandar Seri Begawan (cap.), Brunei	E 4	36
Bandelier N.M., N.Mex.	C 3	94
Bandjarmasin, Indon.	E 6	36
Bandung, Indon.	D 7	36
Banff, Scot.	E 2	23
Banff N.P., Alta.	B 4	58
Bangalore, India	D 6	34
Bangkok (cap.), Thai.	D 4	35
Bangladesh	F-G 4	34
Bangor, Maine	F 6	82
Bangor, Pa.	M 4	101
Bangui (cap.), C.A.R.	K11	38
Bangweulu (lake), Zambia	N14	39
Banjul (cap.), Gambia	C 9	38
Banks (isl.), N.W.T.	F 2	60
Banning, Calif.	J10	69
Bannockburn, Scot.	B 1	23
Baraboo, Wis.	G 9	111
Barbados	G 4	45
Barberton, Ohio	G 4	98
Barcelona, Spain	H 2	27
Barcelona, Ven.	G 2	45
Bardstown, Ky.	F 5	80
Bareilly, India	D 3	34
Barents (sea), Europe	D 2	30
Bar Harbor, Maine	G 7	82
Bari, Italy	F 4	28
Barnaul, U.S.S.R.	J 4	30
Barnstable, Mass.	N 6	84
Baroda, India	C 4	34
Barotseland (reg.); Zambia	L15	39
Barquisimeto, Ven.	F 2	45
Barranquilla, Col.	F 1	42
Barre, Vt.	C 2	107
Barrie, Ont.	E 3	55
Barrington, Ill.	E 1	76
Barrington, N.J.	B 3	93
Barrington, R.I.	J 6	84
Barrow (pt.), Alaska	G 1	66
Barstow, Calif.	H 9	69
Bartlesville, Okla.	K 1	99
Bartonville, Ill.	D 3	76
Bartow, Fla.	E 4	72
Basel, Switz.	C 1	27
Bashkir A.S.S.R., U.S.S.R.	F 4	30
Basilicata (reg.), Italy	E 4	28
Basra, Iraq	E 3	33
Basse-Terre (cap.), Guad.	F 4	45
Basseterre (cap.), St. C.-N.-A.	F 3	45
Bastogne, Belg.	G 7	24
Bastrop, La.	G 1	81
Batavia, Ill.	E 2	76
Batavia, N.Y.	D 5	95
Batesville, Ark.	G 2	68
Bath, Eng.	E 5	23
Bath, Maine	D 8	82
Bath, N.Y.	F 6	95
Baton Rouge (cap.), La.	K 2	81
Battle Creek, Mich.	D 6	85
Batumi, U.S.S.R.	E 5	30
Bavaria (state), W.Ger.	D 4	25
Bayamón, P.R.	G 1	45
Bay City, Mich.	F 5	85
Bay City, Tex.	H 9	105
Bayeux, France	C 3	26
Baykal (lake), U.S.S.R.	L 4	30

Name	Index Ref.	Plate No.
Danville, Ind.	D 5	77
Danville, Pa.	J 4	101
Danville, Va.	E 7	108
Danzig (gulf), Pol.	D 1	31
Darby, Pa.	M 7	101
Dardanelles (str.), Turkey	A 2	33
Dar es Salaam (cap.), Tanz.	O13	39
Darfur (prov.), Sudan	L 9	38
Darien, Conn.	B 4	71
Darién (mts.), Pan.	E 3	46
Darjeeling, India	F 3	34
Darlac (plat.), Viet.	F 4	35
Darling (riv.), Aust.	G 6	40
Darlington, S.C.	H 3	102
Darmstadt, W.Ger.	C 4	25
Dartford, Eng.	C 5	23
Dartmouth, Mass.	K 6	84
Dartmouth, N.S.	E 4	51
Daru, P.N.G.	B 7	36
Darwin, Aust.	E 2	40
Dauphin, Man.	B 3	56
Davao, Phil.	H 4	36
Davenport, Iowa	M 5	78
David, Pan.	D 3	46
Davie, Fla.	B 4	72
Davis (sea), Ant.	C 5	20
Davis, Calif.	B 8	69
Davis (str.), N.A.	M 3	60
Davison, Mich.	F 5	85
Davos, Switz.	E 2	27
Dawson, Ga.	D 7	73
Dawson, Yukon	E 3	60
Dawson Creek, B.C.	G 2	59
Dayton, Ky.	L 1	80
Dayton, Ohio	B 6	98
Dayton, Tenn.	L 3	104
Daytona Beach, Fla.	F 2	72
Dead (sea), Asia	C 4	31
Deadwood, S.Dak.	B 5	103
Dearborn, Mich.	B 7	85
Dearborn Hts., Mich.	B 7	85
Death Valley (depr.), Calif.	H 7	69
Death Valley N.M., U.S.	H-J 7	69
Deauville, France	C 3	26
Debrecen, Hung.	F 3	29
Decatur, Ala.	D 1	65
Decatur, Ga.	D 3	73
Decatur, Ill.	E 4	76
Decatur, Ind.	H 3	77
Deccan (plat.), India	D 5-6	34
Decorah, Iowa	K 2	78
Dedham, Mass.	C 7	84
Dee (riv.), Scot.	E 2	23
Deerfield, Ill.	F 1	76
Deerfield Bch., Fla.	F 5	72
Deer Park, N.Y.	E 2	95
Deer Park, Ohio	C 9	98
Deer Park, Tex.	K 2	105
Defiance, Ohio	B 3	98
Dehra Dun, India	D 2	34
De Kalb, Ill.	E 2	76
Delagoa (bay), Moz.	N17	39
De Land, Fla.	E 2	72
Delano, Calif.	F 8	69
Delavan, Wis.	J10	111
Delaware, Ohio	E 5	98
Delaware (bay), U.S.	C 5	93
Delaware (riv.), U.S.	B 5	93
Delaware (state), U.S.		83
Delaware Water Gap N.R.A., U.S.	C 1	93
Del City, Okla.	G 4	99
Delft, Neth.	D 3	24
Delgado (cape), Moz.	P14	39
Delhi, India	D 3	34
Deloraine, Man.	B 5	56
Delphos, Ohio	B 4	98
Delran, N.J.	B 3	93
Delray Beach, Fla.	F 5	72
Del Rio, Tex.	D 8	105
Del Rosa, Calif.	E10	69
Demarest, N.J.	C 1	93
Demavend (mt.), Iran	F 2	33
Deming, N.Mex.	B 6	94
Demopolis, Ala.	C 6	65
Denbigh, Wales	E 4	23
Denham Sprs., La.	L 1	81
Denison, Iowa	C 4	78
Denison, Tex.	H 4	105
Denmark	B 3	24
Denmark (str.)	C11	20
Dennis, Mass.	O 5	84
Denton, Tex.	G 4	105
Denver (cap.), Colo.	K 3	70
Denville, N.J.	E 2	93
De Pere, Wis.	K 7	111
Depew, N.Y.	C 5	95
Deptford, N.J.	B 4	93
Derby, Aust.	C 3	40
Derby, Conn.	C 3	71
Derby, Eng.	F 4	23
Derby, Kans.	E 4	79
De Ridder, La.	D 5	81
Derna, Libya	L 5	38
Derry, N.H.	D 6	92
Des Moines (cap.), Iowa	G 5	78
De Soto, Mo.	M 6	88
De Soto, Tex.	G 2	105
De Soto N.Mem., Fla.	D 4	72
Des Peres, Mo.	O 3	88
Des Plaines, Ill.	A 1	76
Dessau, E.Ger.	E 3	25
Dessye, Eth.	O 9	38
Detroit, Mich.	B 7	85
Detroit Lakes, Minn.	C 4	86
Deva, Rum.	C 2	29
Deventer, Neth.	H 3	24
Devils (isl.), Fr.Gui.	K 2	42
Devils Lake, N.Dak.	N 3	97
Devils Postpile N.M., Calif.	F 6	69
Devils Tower N.M., Wyo.	H 1	112
Devon, Alta.	D 3	58
Devon (isl.), N.W.T.	K 2	60
Dexter, Mo.	N 9	88
Dezhnev (cape), U.S.S.R.	T 3	30
Dhahran, Saudi Ar.	E 4	33
Dhaulagiri (mt.), Nepal	E 3	34
Dhofar (reg.), Oman	F 6	33
Diamond (head), Hawaii	F 2	74
Dickinson, N.Dak.	E 6	97
Dickson, Tenn.	G 2	104
Dickson City, Pa.	L 3	101
Diego Garcia (isls.), B.I.O.T.	L10	32
Dien Bien Phu, Viet.	D 2	35
Dieppe, France	D 3	26
Dieppe, N.Br.	F 2	52
Digby, N.S.	C 4	51
Dijon, France	F 4	26
Dili, Indonesia	H 7	36
Dillon, S.C.	J 3	102
Dinaric Alps (mts.), Yugo.	B 3	29
Dinosaur N.M., U.S.	A 1	70
Dinuba, Calif.	F 7	69
Diomede (isls.)	E 1	66
Dire Dawa, Eth.	P10	38
Disappointment (cape), Wash.	A 4	109
District Hts., Md.	C 5	83
District of Columbia, U.S.	B 5	83
Dixon, Ill.	D 2	76
Dixon Entrance (str.), N.A.	M 2	66
Diyarbakir, Turkey	C 2	33
Djajapura, Indon.	K 6	36
Djakarta (cap.), Indon.	D 7	36
Djerba (isl.), Tun.	J 5	38
Djibouti (cap.), Djibouti	P 9	38
Djokjakarta, Indon.	D 7	36
Dnepropetrovsk, U.S.S.R.	D 5	30
Dnieper (riv.), U.S.S.R.	D 5	30
Dniester (riv.), U.S.S.R.	C 5	30
Dobbs Ferry, N.Y.	C 2	95
Dodecanese (isls.), Greece	D 5	29
Dodge City, Kans.	B 4	79
Dodoma, Tanz.	O13	39
Doha (cap.), Qatar	F 4	33
Dolgellau, Wales	E 4	23
Dolomite Alps (mts.), Italy	C 1	28
Dolton, Ill.	B 2	76
Dominica	G 4	45
Dominican Republic	D 3	45
Domrémy-la-Pucelle, France	F 3	26
Don (riv.), Eng.	F 4	23
Don (riv.), Ont.	J 4	55
Don (riv.), Scot.	E 2	23
Don (riv.), U.S.S.R.	E 5	30
Donaldsonville, La.	K 3	81
Doncaster, Eng.	F 4	23
Dondra (head), Sri Lanka	E 7	34
Donegal, Ire.	B 3	23
Donets (riv.), U.S.S.R.	D 5	30
Donetsk, U.S.S.R.	D 5	30
Dongola, Sudan	M 8	38
Donna, Tex.	F11	105
Donora, Pa.	C 5	101
Doorn, Neth.	F 3	24
Doraville, Ga.	D 3	73
Dorchester, Mass.	D 7	84
Dorchester Hts. N.H.S., Mass.	D 7	84
Dordogne (riv.), France	D 5	26
Dordrecht, Neth.	E 4	24
Dormont, Pa.	B 7	101
Dornbirn, Aus.	A 3	29
Dortmund, W.Ger.	B 3	25
Dothan, Ala.	H 8	65
Douai, France	E 2	26
Douala, Cameroon	J11	38
Douglas, Ariz.	F 7	67
Douglas, Ga.	G 7	73
Douglas (cap.), I. of Man	D 3	23
Douglasville, Ga.	C 3	73
Douro (Duero) (riv.), Europe	C 2	27
Dover (cap.), Del.	M 4	83
Dover, Eng.	G 5	23
Dover (str.), Europe	G 5	23
Dover, N.H.	E 5	92
Dover, N.J.	D 2	93
Dover, Ohio	G 4	98
Dovrefjell (mts.), Nor.	B 2	24
Dowagiac, Mich.	D 6	85
Downers Grove, Ill.	A 2	76
Downey, Calif.	C11	69
Downingtown, Pa.	L 5	101
Downpatrick, N.Ire.	C 3	23
Doylestown, Pa.	M 5	101
Dracut, Mass.	J 2	84
Dragons Mouth (passage)	F 5	45
Drake (passage)	C15	20
Dráma, Greece	C 3	29
Drava (riv.), Europe	D 4	29
Dresden, E.Ger.	E 3	25
Drumheller, Alta.	D 4	58
Duarte, Calif.	D10	69
Dubai, U.A.E.	F 4	33
Dubbo, Aust.	H 6	40
Dublin, Ga.	G 5	73
Dublin (cap.), Ire.	C 4	23
Du Bois, Pa.	E 3	101
Dubrovnik, Yugo.	B 3	29
Dubuque, Iowa	M 3	78
Dudley, Mass.	G 4	84
Duero (Douro) (riv.), Europe	D 2	27
Dufourspitze (mt.), Switz.	C 3	27
Duisburg, W.Ger.	B 3	25
Duluth, Minn.	F 4	86
Dumas, Tex.	C 2	105
Dumbarton, Scot.	A 1	23
Dumfries, Scot.	E 3	23
Dumont, N.J.	C 1	93
Dunbar, W.Va.	C 4	110
Duncan, Okla.	G 5	99
Duncanville, Tex.	G 2	105
Dundalk, Ire.	C 4	23
Dundalk, Md.	H 3	83
Dundas, Ont.	D 4	55
Dundee, Scot.	E 2	23
Dunedin, Fla.	B 2	72
Dunedin, N.Z.	L 7	40
Dunellen, N.J.	D 2	93
Dunfermline, Scot.	D 2	23
Dunkirk (Dunkerque), France	E 2	26
Dunkirk, N.Y.	B 5	95
Dún Laoghaire, Ire.	C 4	23
Dunmore, Pa.	L 3	101
Dunn, N.C.	H 4	96
Duquesne, Pa.	C 7	101
Du Quoin, Ill.	D 5	76
Durango, Colo.	D 8	70
Durango, Mex.	D 3	46
Durant, Okla.	K 6	99
Durazno, Urug.	J10	43
Durban, S.Afr.	N17	39
Durham, Eng.	F 3	23
Durham, N.H.	E 5	92
Durham, N.C.	H 2	96
Durrës, Alb.	B 3	29
Duryea, Pa.	L 3	101
Dushanbe, U.S.S.R.	G 6	30
Düsseldorf, W.Ger.	B 3	25
Dutch Harbor, Alaska	E 4	66
Duxbury, Mass.	M 4	84
Dvina, No. (riv.), U.S.S.R.	E 3	30
Dvina, Western (riv.), U.S.S.R.	C 4	30
Dzaoudzi (cap.), Mayotte	R14	39
Dzerzhinsk, U.S.S.R.	E 4	30
Dzhambul, U.S.S.R.	H 5	30
Dzungaria (reg.), China	C 3	37

E

Name	Index Ref.	Plate No.
Eagle Pass, Tex.	D 9	105
Ealing, Eng.	B 5	23
Easley, S.C.	B 2	102
East Alton, Ill.	B 6	76
E. Aurora, N.Y.	C 5	95
Eastbourne, Eng.	G 5	23
E. Bridgewater, Mass.	L 4	84
E. Brunswick, N.J.	E 3	93
E. Chicago, Ind.	C 1	77
E. Chicago Hts., Ill.	B 3	76

Name	Index Ref.	Plate No.
Falmouth, Maine	C 8	82
Falmouth, Mass.	M 6	84
Falun, Sweden	C 2	24
Famagusta, Cyprus	B 3	33
Famatina (mts.), Arg.	G 9	43
Fanning (isl.), Kiribati	L 5	41
Fanwood, N.J.	E 2	93
Farallon (isls.), Calif.	B 6	69
Farewell (cape), Greenl.	O 4	44
Fargo, N.Dak.	S 6	97
Faribault, Minn.	E 6	86
Farmers Branch, Tex.	G 1	105
Farmingdale, N.Y.	D 2	95
Farmington, Conn.	D 2	71
Farmington, Maine	C 6	82
Farmington, Mich.	F 6	85
Farmington, Mo.	M 7	88
Farmington, N.Mex.	A 2	94
Faro, Port.	B 4	27
Farrell, Pa.	A 3	101
Fatehgarh, India	D 3	34
Fatshan, China	H 7	37
Fayetteville, Ark.	B 1	68
Fayetteville, N.C.	H 4	96
Fayetteville, Tenn.	H 4	104
Fear (cape), N.C.	K 7	96
Feeding Hills, Mass.	D 4	84
Fehmarn (isl.), W.Ger.	D 1	25
Feldkirch, Aus.	A 3	29
Fénérive, Mad.	R15	39
Fenton, Mich.	F 6	85
Fergana, U.S.S.R.	H 5	30
Fergus Falls, Minn.	B 4	86
Ferguson, Mo.	P 2	88
Fernandina Bch., Fla.	E 1	72
Fernando Po (Bioko) (isl.), Eq. Guin.	H11	38
Fernie, B.C.	K 5	59
Ferozepore, India	C 2	34
Ferrara, Italy	C 2	28
Festus, Mo.	M 6	88
Fez, Mor.	F 5	38
Fezzan (reg.), Libya	J 6	38
Ffestiniog, Wales	E 4	23
Fianarantsoa, Mad.	R16	39
Fichtelgebirge (mts.), W.Ger.	D 3	25
Fifth Cataract (falls), Sudan	N 8	38
Fiji	H 8	41
Fillmore, Calif.	G 9	69
Findlay, Ohio	C 3	98
Finisterre (cape), Spain	B 1	27
Finland	E 1-2	24
Finland (gulf), Europe	E 3	24
Finlay (riv.), B.C.	E 1	59
Finsteraarhorn (mt.), Switz.	D 2	27
Fircrest, Wash.	C 3	109
Fire I. N.Sea., N.Y.	F 2	95
Firenze (Florence), Italy	C 3	28
Fitchburg, Mass.	G 2	84
Fitzgerald, Ga.	F 7	73
Fiumicino, Italy	C 4	28
Flagstaff, Ariz.	D 3	67
Flaming Gorge (res.), U.S.	C 4	112
Flanders (prov's), Belg.	B-C 5-6	24
Flathead (lake), Mont.	C 3	89
Flat Rock, Mich.	F 6	85
Flattery (cape), Wash.	A 2	109
Flatwoods, Ky.	M 4	80
Flemington, N.J.	D 2	93
Flensburg, W.Ger.	C 1	25
Flin Flon, Man.	H 3	56
Flint, Mich.	F 6	85
Flora, Ill.	E 5	76
Floral Park, N.Y.	A 2	95
Florence, Ala.	C 1	65
Florence, Italy	C 3	28
Florence, Ky.	J 2	80
Florence, S.C.	H 3	102
Flores (isl.), Indon.	G 7	36
Florham Park, N.J.	E 2	93
Florianópolis, Braz.	L 9	43
Florida (bay), Fla.	F 7	72
Florida (straits), N.A.	B 1	45
Florida (state), U.S.		72
Florida, Urug.	J10	43
Florida City, Fla.	F 6	72
Florissant, Mo.	P 2	88
Florissant Fossil Beds N.M., Colo.	J 5	70
Flossmoor, Ill.	B 3	76
Flushing, Mich.	F 5	85
Flushing, Neth.	B 5	24
Fly (riv.), P.N.G.	B 7	36
Foggia, Italy	E 4	28
Foix, France	D 6	26
Folcroft, Pa.	M 7	101
Folkestone, Eng.	G 5	23
Fond du Lac, Wis.	K 8	111
Fongafale (cap.), Tuvalu	H 6	41
Fonseca (gulf), C.A.	C 2	46
Fontainebleau, France	E 3	26
Fontana, Calif.	E10	69
Foochow, China	J 6	37
Forest Acres, S.C.	F 3	102
Forest City, N.C.	F 8	96
Forest Grove, Oreg.	A 2	100
Forest Hill, Tex.	F 2	105
Forest Hills, Pa.	C 7	101
Forest Park, Ga.	D 3	73
Forest Park, Ill.	B 2	76
Forest Park, Ohio	B 9	98
Forfar, Scot.	E 2	23
Forlì, Italy	D 2	28
Formosa, Arg.	J 9	43
Formosa (Taiwan) (isl.), China	K 7	37
Forrest City, Ark.	J 3	68
Fortaleza, Braz.	N 4	42
Fort Atkinson, Wis.	J10	111
Ft. Belvoir, Va.	K 3	108
Ft. Benning, Ga.	B 6	73
Ft. Bliss, Tex.	A10	105
Ft. Bowie N.H.S., Ariz.	F 6	67
Ft. Bragg, N.C.	M 4	96
Ft. Campbell, U.S.	B 7	80
Ft. Caroline N.Mem., Fla.	E 1	72
Ft. Clatsop N.Mem., Oreg.	C 1	100
Ft. Collins, Colo.	J 1	70
Ft. Davis N.H.S., Tex.	D11	105
Ft-de-France (cap.), Mart.	G 4	45
Ft. Dodge, Iowa	E 3	78
Ft. Donelson N.M.P., Tenn.	F 8	104
Ft. Erie, Ont.	E 5	55
Ft. Frederica N.M., Ga.	K 8	73
Ft. George G. Meade, Md.	H 4	83
Forth (firth), Scot.	E 2	23
Fort Jefferson N.M., Fla.	C 7	72
Ft. Knox, Ky.	F 5	80
Ft. Laramie N.H.S., Wyo.	H 3	112
Ft. Larned N.H.S., Kans.	C 3	79
Ft. Lauderdale, Fla.	C 4	72
Ft. Lee, N.J.	C 2	93
Ft. Lee, Va.	K 6	108
Ft. Leonard Wood, Mo.	H 7	88
Ft. Macleod, Alta.	D 5.	58
Ft. Madison, Iowa	L 7	78
Ft. Matanzas N.M., Fla.	E 2	72
Ft. McHenry N.M., Md.	H 3	83
Ft. McMurray, Alta.	E 1	58
Ft. Mitchell, Ky.	K 2	80
Ft. Morgan, Colo.	M 2	70
Ft. Myers, Fla.	E 5	72
Ft. Necessity N.B.P., Pa.	C 6	101
Ft. Payne, Ala.	G 2	65
Ft. Peck Lake, (res.), Mont.	K 3	89
Ft. Pierce, Fla.	F 4	72
Ft. Point N.H.S., Calif.	J 2	69
Ft. Providence, N.W.T.	G 3	60
Ft. Pulaski N.M., Ga.	L 6	73
Ft. Raleigh N.H.S., N.C.	T 3	96
Ft. Richardson, Alaska	C 1	66
Ft. Riley, Kans.	F 2	79
Ft. Scott, Kans.	H 4	79
Ft. Sill, Okla.	F 5	99
Ft. Simpson, N.W.T.	F 3	60
Ft. Smith, Ark.	B 3	68
Ft. Smith, N.W.T.	G 3	60
Ft. Smith N.H.S., Ark.	B 3	68
Ft. Stanwix N.M., N.Y.	J 4	95
Ft. Stockton, Tex.	A 7	105
Ft. Sumter N.M., S.C.	H 6	102
Ft. Thomas, Ky.	L 1	80
Ft. Ticonderoga, N.Y.	O 3	95
Ft. Union N.M., N.Mex.	E 3	94
Ft. Union Trading Post N.H.S., N.Dak.	B 3	97
Ft. Valley, Ga.	E 5	73
Ft. Vancouver N.H.S., Wash.	C 5	109
Ft. Walton Bch., Fla.	C 6	72
Ft. Wayne, Ind.	G 2	77
Ft. William, Scot.	D 2	23
Ft. Worth, Tex.	E 2	105
Forty Fort, Pa.	L 3	101
Fostoria, Ohio	D 3	98
Fountain Hill, Pa.	L 4	101
Fountain Valley, Calif.	D11	69
Fourth Cataract (falls), Sudan	N 8	38
Foxboro, Mass.	J 4	84
Foxe (basin), N.W.T.	K 3	60
Fox Point, Wis.	M 1	111
Frackville, Pa.	K 4	101
Framingham, Mass.	A 7	84
France		26
Francis Case (lake), S.Dak.	L-M 6-7	103
Francistown, Botswana	M16	39
Frankfort, Ind.	E 4	77
Frankfort (cap.), Ky.	G 4	80
Frankfurt-am-Main, W.Ger.	C 3	25
Frankfurt-an-der-Oder, E.Ger.	F 2	25
Franklin, Ind.	E 6	77
Franklin, Ky.	D 7	80
Franklin, La.	G 7	81
Franklin, Mass.	J 4	84
Franklin, N.H.	C 5	92
Franklin, N.J.	D 3	93
Franklin (dist.), N.W.T.	G-L 2	60
Franklin, Ohio	B 6	98
Franklin, Pa.	C 3	101
Franklin, Tenn.	H 3	104
Franklin, Va.	L 7	108
Franklin, Wis.	L 2	111
Franklin D. Roosevelt (lake), Wash.	G 2	109
Franklin Lakes, N.J.	B 1	93
Franklin Park, Ill.	A 2	76
Franz Josef Land (isls.), U.S.S.R.	E-G 1	30
Fraser (riv.), B.C.	F 4	59
Fraser, Mich.	B 6	85
Fredericia, Den.	B 3	24
Frederick, Md.	E 3	83
Frederick, Okla.	D 6	99
Fredericksburg, Tex.	E 7	105
Fredericksburg, Va.	J 4	108
Fredericton (cap.), N.Br.	D 3	52
Fredonia, N.Y.	B 6	95
Freehold, N.J.	E 3	93
Freeport, Ill.	D 1	76
Freeport, N.Y.	B 3	95
Freeport, Tex.	J 9	105
Freetown (cap.), S.Leone	D10	38
Freiburg, W.Ger.	B 5	25
Freising, W.Ger.	D 4	25
Fremantle, Aust.	B 2	40
Fremont, Calif.	K 3	69
Fremont, Nebr.	H 3	90
Fremont, Ohio	D 3	98
French Guiana	K 3	42
Frenchman (riv.), N.A.	E 1	61
French Polynesia	L-N 8	41
Fresno, Calif.	F 7	69
Fribourg, Switz.	C 2	27
Fridley, Minn.	G 5	86
Friedrichshafen, W.Ger.	C 5	25
Friendswood, Tex.	J 2	105
Friesland (prov.), Neth.	G 1	24
Frio (cape), Braz.	M 8	43
Friuli-Venezia Giulia (reg.), Italy	D 1	28
Frobisher Bay, N.W.T.	M 3	60
Front Royal, Va.	H 3	108
Frostburg, Md.	C 7	83
Frunze, U.S.S.R.	H 5	30
Fuerteventura (isl.), Spain	C 4	27
Fujaira, U.A.E.	G 4	33
Fuji (mt.), Japan	E 4	36
Fukien (prov.), China	J 6	37
Fukui, Japan	D 3	36
Fukuoka, Japan	C 4	36
Fukushima, Japan	F 3	36
Fulda, W.Ger.	C 3	25
Fullerton, Calif.	D11	69
Fulton, Mo.	J 5	88
Fulton, N.Y.	H 4	95
Fultondale, Ala.	E 3	65
Funchal, Port.	A 2	27
Fundy (bay), N.A.	E 3	52
Fundy N.P., N.Br.	E 3	52
Fürstenfeldbruck, W.Ger.	D 4	25
Fürth, W.Ger.	D 4	25
Fushun, China	K 3	37
Füssen, W.Ger.	D 5	25

G

Name	Index Ref.	Plate No.
Gabès (gulf), Tun.	J 5	38
Gabon	J12	39
Gaborone (cap.), Botswana	L16	39
Gadsden, Ala.	G 2	65
Gaeta, Italy	D 4	28
Gaffney, S.C.	D 1	102
Gafsa, Tun.	H 5	38
Gahanna, Ohio	E 5	98
Gainesville, Fla.	D 2	72
Gainesville, Ga.	E 2	73

	Index Ref.	Plate No.
Kénitra, Mor.	D 5	38
Kenmare, Ire.	B 5	23
Kenmore, N.Y.	C 5	95
Kennebunk, Maine	B 9	82
Kenner, La.	O 4	81
Kennesaw Mtn. N.B.P., Ga.	C 3	73
Kennett, Mo.	M10	88
Kennewick, Wash.	F 4	109
Kenora, Ont.	B 3	54
Kenosha, Wis.	M 3	111
Kent, Ohio	H 3	98
Kent, Wash.	C 3	109
Kenton, Ohio	C 4	98
Kentucky (lake), U.S.	J 3	61
Kentucky (state), U.S.		80
Kentville, N.S.	D 3	51
Kentwood, Mich.	D 6	85
Kenya	O11	39
Kenya (mt.), Kenya	O12	39
Keokuk, Iowa	L 8	78
Kerala (state), India	D 6	34
Kerguélen (isls.)	N 8	19
Kérkira (isl.), Greece	B 4	29
Kermadec (isls.), N.Z.	J 9	40
Kerman, Iran	G 3	33
Kermanshah, Iran	E 3	33
Kermit, Tex.	B 6	105
Kerrville, Tex.	E 7	105
Kerulen (riv.), Asia	H 2	37
Keta, Ghana	G10	38
Ketchikan, Alaska	N 2	66
Kettering, Ohio	B 6	98
Kewanee, Ill.	C 2	76
Keweenaw (pt.), Mich.	B 1	85
Keyport, N.J.	E 3	93
Keyser, W.Va.	J 2	110
Key West, Fla.	E 7	72
Khabarovsk, U.S.S.R.	O 5	30
Khaniá, Greece	C 5	29
Kharagpur, India	F 4	34
Khârga (oasis), Egypt	N 6	38
Khar'kov, U.S.S.R.	D 4	30
Khartoum (cap.), Sudan	N 8	38
Khemmarat, Thai.	E 4	35
Kherson, U.S.S.R.	D 5	30
Khíos (isl.), Greece	D 4	29
Khorramshahr, Iran	E 3	33
Khotan, China	A 4	37
Khulna, Bang.	F 4	34
Khyber (pass), Pak.	C 2	34
Kiangsi (prov.), China	J 6	37
Kiangsu (prov.), China	K 5	37
Kiel, W.Ger.	D 1	25
Kiel (canal), W.Ger.	C 1	25
Kielce, Pol.	E 3	31
Kiev, U.S.S.R.	D 4	30
Kigali (cap.), Rwanda	N12	39
Kigoma, Tanz.	N12	39
Kilauea (crater), Hawaii	H 6	74
Kildare, Ire.	C 4	23
Kilgore, Tex.	K 5	105
Kilimanjaro (mt.), Tanz.	O12	39
Kilkenny, Ire.	C 4	23
Killarney, Ire.	B 4	23
Killeen, Tex.	G 6	105
Killingly, Conn.	H 1	71
Kimberley (plat.), Aust.	D 3	40
Kimberley, B.C.	K 5	59
Kimberley, S.Afr.	L17	39
Kimberly, Wis.	K 7	111
Kinabalu (mt.), Malaysia	F 4	36
Kincardine, Scot.	B 1	23
Kindersley, Sask.	B 4	57
King (isl.), Aust.	G 7	40
Kingman, Ariz.	A 3	67

	Index Ref.	Plate No.
Kings Canyon N.P., Calif.	G 7	69
Kingsford, Mich.	A 3	85
Kings Mtn., N.C.	C 4	96
Kings Mtn. N.M.P., S.C.	E 1	102
Kings Point, N.Y.	A 2	95
Kingsport, Tenn.	Q 1	104
Kingston (cap.), Jam.	C 3	45
Kingston, Mass.	M 5	84
Kingston, N.Y.	M 7	95
Kingston (cap.), Norfolk I.	G 8	41
Kingston, Ont.	H 3	55
Kingston, Pa.	K 3	101
Kingstown (cap.), St. Vincent & Grens.	G 4	45
Kingsville, Ont.	B 5	55
Kingsville, Tex.	G10	105
Kinloch, Mo.	P 2	88
Kinnelon, N.J.	E 2	93
Kinross, Scot.	E 2	23
Kinsale, Ire.	B 5	23
Kinshasa (cap.), Zaire	K12	39
Kioga (lake), Uganda	N11	39
Kirgiz S.S.R., U.S.S.R.	H 5	30
Kiribati	J 6	41
Kirin, China	L 3	37
Kirkcaldy, Scot.	C 1	23
Kirkcudbright, Scot.	E 3	23
Kirkenes, Nor.	E 1	24
Kirkland, Wash.	B 2	109
Kirkland Lake, Ont.	D 3	54
Kirksville, Mo.	H 2	88
Kirkuk, Iraq	D 2	33
Kirkwall, Scot.	E 1	23
Kirkwood, Mo.	O 3	88
Kirov, U.S.S.R.	E 4	30
Kirovabad, U.S.S.R.	E 5	30
Kirovograd, U.S.S.R.	D 5	30
Kirtland, Ohio	H 2	98
Kiruna, Sweden	D 1	24
Kisangani, Zaire	M11	39
Kishinev, U.S.S.R.	C 5	30
Kiska (isl.), Alaska	J 4	66
Kismayu, Somalia	P12	39
Kissimmee, Fla.	E 3	72
Kistna (riv.), India	D 5	34
Kita Iwo (isl.), Japan	D 3	41
Kitakyushu, Japan	C 4	36
Kitchener, Ont.	D 4	55
Kitimat, B.C.	C 3	59
Kittanning, Pa.	D 4	101
Kittery, Maine	B 9	82
Kitzbühel, Aus.	B 3	29
Kiungchow (str.), China	G 7	37
Kivu (lake), Afr.	M12	39
Kizel, U.S.S.R.	F 4	30
Kızılırmak (riv.), Turkey	B 1	33
Kjölen (mts.), Europe	C 1	24
Kladno, Czech.	C 1	29
Klagenfurt, Aus.	C 3	29
Klaipėda, U.S.S.R.	B 4	30
Klamath Falls, Oreg.	F 5	100
Kleve, W.Ger.	B 3	25
Klondike (riv.), Yukon	E 3	60
Kluane (lake), Yukon	E 3	60
Knoxville, Iowa	G 6	78
Knoxville, Tenn.	O 3	104
Kobdo, Mong.	D 2	37
Kobe, Japan	E 4	36
Koblenz, W.Ger.	B 3	25
Kobuk (riv.), Alaska	G 1	66
Kodiak (isl.), Alaska	H 3	66
Kokomo, Ind.	E 4	77
Koko Nor (lake), China	E 4	37
Kola (pen.), U.S.S.R.	D 3	30
Kolar Gold Fields, India	D 6	34

	Index Ref.	Plate No.
Kolguyev (isl.), U.S.S.R.	E 3	30
Kolhapur, India	C 5	34
Köln (Cologne), W.Ger.	B 3	25
Kołobrzeg, Pol.	B 1	31
Kolyma (range), U.S.S.R.	Q 3	30
Komárno, Czech.	E 3	29
Kompong Chhnang, Camb.	D 4	35
Kompong Thom, Camb.	E 4	35
Komsomol'sk, U.S.S.R.	O 4	30
Kongsberg, Nor.	B 3	24
Königssee (lake), W.Ger.	E 5	25
Konin, Pol.	D 2	31
Köniz, Switz.	C 2	27
Konstanz, W.Ger.	C 5	25
Kontum (plat.), Viet.	E 4	35
Konya, Turkey	B 2	33
Kootenay (riv.), B.C.	K 5	59
Koper, Yugo.	A 2	29
Korçë, Alb.	C 3	29
Korčula (isl.), Yugo.	B 3	29
Kordofan (prov.), Sudan	M 9	38
Korea		36
Koror, T.T.P.I.	D 5	41
Korsakov, U.S.S.R.	P 5	30
Koryak (range), U.S.S.R.	R 3	30
Kos (isl.), Greece	D 4	29
Kosciusko (mt.), Aust.	H 7	40
Kosciusko, Miss.	E 4	87
Košice, Czech.	F 2	29
Kostroma, U.S.S.R.	E 4	30
Koszalin, Pol.	C 1	31
Kota Bharu, Malaysia	D 6	35
Kota Kinabalu, Malaysia	E 4	36
Kotka, Fin.	E 2	24
Kotor, Yugo.	B 3	29
Kowloon, Hong Kong	H 7	37
Koyukuk (riv.), Alaska	G 1	66
Kozhikode, India	D 6	34
Kra (isth.), Thai.	C 5	35
Kragujevac, Yugo.	C 2	29
Krakatau (isl.), Indon.	C 7	36
Krasnodar, U.S.S.R.	D 5	30
Krasnovodsk, U.S.S.R.	F 5	30
Krasnoyarsk, U.S.S.R.	K 4	30
Krefeld, W.Ger.	B 3	25
Kremenchug, U.S.S.R.	D 5	30
Krishna (Kistna) (riv.), India	D 5	34
Kristiansand, Nor.	B 3	24
Kristiansund, Nor.	B 2	24
Kristinehamn, Sweden	C 3	24
Kristinestad, Fin.	D 2	24
Krivoy Rog, U.S.S.R.	D 5	30
Krk (isl.), Yugo.	A 2	29
Kuala Lumpur (cap.), Malaysia	D 7	35
Kuching, Malaysia	D 5	36
Kufra (oasis), Libya	L 7	38
Kuldja, China	B 3	37
Kuma (riv.), U.S.S.R.	E 5	30
Kumamoto, Japan	C 4	36
Kumanovo, Yugo.	C 3	29
Kumasi, Ghana	F10	38
Kunlun (mts.), Asia	B-D 4	37
Kunming, China	F 6	37
Kuopio, Fin.	E 2	24
Kupang, Indon.	G 8	36
Kura (riv.), U.S.S.R.	E 6	30
Kurdistan (reg.), Asia	D 2	33
Kürdzhali, Bulg.	D 3	29
Kure, Japan	D 4	36
Kurgan, U.S.S.R.	G 4	30
Kuril (isls.), U.S.S.R.	P 5	30
Kursk, U.S.S.R.	D 4	30

	Index Ref.	Plate No.
Kuskokwim (riv.), Alaska	G 2	66
Kutaisi, U.S.S.R.	E 5	30
Kutch, Rann of (salt marsh), India	B-C 4	34
Kutztown, Pa.	L 4	101
Kuusamo, Fin.	E 2	24
Kuwait	E 4	33
Kuybyshev, U.S.S.R.	F 4	30
Kwajalein (atoll), T.T.P.I.	G 5	41
Kwangju, S.Korea	B 4	36
Kwangtung (prov.), China	H 7	37
Kweichow (prov.), China	G 6	37
Kweiyang, China	G 6	37
Kwinana, Aust.	B 2	40
Kyaukpyu, Burma	B 3	35
Kyoto, Japan	E 4	36
Kyushu (isl.), Japan	D 4	36
Kyustendil, Bulg.	C 3	29
Kyzyl, U.S.S.R.	K 4	30
Kyzyl-Kum (des.), U.S.S.R.	G 5	30

L

	Index Ref.	Plate No.
Laayoune, Morocco	D 6	38
Labrador (dist.), Newf.	B-C 2-3	50
Labuan (isl.), Malaysia	E 4	36
Laccadive (isls.), India	C 6	34
La Ceiba, Hond.	C 2	46
La Chaux-de-Fonds, Switz.	B 1	27
Lachine, Que.	H 4	53
Lackawanna, N.Y.	B 5	95
Lac-Mégantic, Que.	G 4	53
Laconia, N.H.	E 4	92
La Coruña, Spain	B 1	27
La Crosse, Wis.	D 8	111
Ladakh (reg.), India	D 2	34
Ladoga (lake), U.S.S.R.	D 3	30
Ladue, Mo.	P 3	88
Ladysmith, S.Afr.	N17	39
Lae, P.N.G.	B 7	36
Lafayette, Calif.	K 2	69
La Fayette, Ga.	B 1	73
Lafayette, Ind.	D 4	77
Lafayette, La.	F 6	81
La Follette, Tenn.	N 2	104
Lagos (cap.), Nig.	G10	38
La Grande, Oreg.	J 2	100
La Grange, Ga.	B 4	73
La Grange, Ill.	A 2	76
La Grange Park, Ill.	A 2	76
La Guaira, Ven.	G 1	42
Laguna Beach, Calif.	G10	69
La Habra, Calif.	D11	69
Lahore, Pak.	C 2	34
Lahti, Fin.	E 2	24
La Junta, Colo.	M 7	70
Lake Bluff, Ill.	F 1	76
Lake Charles, La.	D 6	81
Lake Chelan N.R.A., Wash.	E 2	109
Lake City, Fla.	D 1	72
Lake City, S.C.	H 4	102
Lake Forest, Ill.	F 1	76
Lake Havasu City, Ariz.	A 4	67
Lakehurst, N.J.	E 3	93
Lake Jackson, Tex.	J 9	105
Lakeland, Fla.	D 3	72
Lake Louise, Alta.	C 4	58
Lake Mead N.R.A., U.S.	G 6	91
Lake of the Woods, N.A.	B 3	54
Lake Oswego, Oreg.	B 2	100
Lake Park, Fla.	F 5	72
Lake Placid, N.Y.	N 2	95
Lake Providence, La.	H 1	81

Name	Index Ref.	Plate No.
Murray, Ky.	D 4	80
Murray, Utah	C 3	106
Murrumbidgee (riv.), Aust.	G 6	40
Murzuk, Libya	J 6	38
Muş, Turkey	D 2	33
Musala (mt.), Bulg.	C 3	29
Musandam, Ras (cape), Oman	G 4	33
Muscat (cap.), Oman	G 5	33
Muscatine, Iowa	L 6	78
Muscle Shoals, Ala.	C 1	65
Musgrave (ranges), Aust.	E 5	40
Muskego, Wis.	K 2	111
Muskegon, Mich.	C 5	85
Muskegon Hts., Mich.	C 5	85
Muskogee, Okla.	M 3	99
Muskoka (lake), Ont.	E 2	55
Musquodoboit (riv.), N.S.	E 4	51
Musselshell (riv.), Mont.	J 3	89
Muzaffarabad, India	B 2	34
Muzaffarpur, India	E 3	34
Muztagh (mt.), China	B 4	37
Muztagh Ata (mt.), China	A 4	37
Mwanza, Tanz.	N12	39
Mweru (lake), Afr.	M13	39
Myitkyina, Burma	C 1	35
Mymensingh, Bang.	G 4	34
Myrtle Beach, S.C.	K 4	102
Mysore, India	D 6	34
Mystic, Conn.	H 3	71

N

Name	Index Ref.	Plate No.
Naas, Ire.	C 4	23
Nablus (Nabulus), Jordan	C 3	31
Nacogdoches, Tex.	J 6	105
Nagaland (state), India	G 3	34
Nagano, Japan	E 3	36
Nagaoka, Japan	E 3	36
Nagasaki, Japan	C 4	36
Nagoya, Japan	E 4	36
Nagpur, India	D 4	34
Nagykanizsa, Hung.	D 3	29
Naha, Japan	G 4	36
Nahariyya, Isr.	C 1	31
Nahuel Huapi (lake), Arg.	F12	43
Nairn, Scot.	E 2	23
Nairobi (cap.), Kenya	O12	39
Naivasha, Kenya	O12	39
Najin, N.Korea	C 2	36
Nakhichevan', U.S.S.R.	E 6	30
Nakhon Ratchasima, Thai.	D 4	35
Nakhon Si Thammarat, Thai.	D 5	35
Nakskov, Den.	**B 3**	**24**
Nakuru, Kenya	O11	39
Namangan, U.S.S.R.	H 5	30
Nam Dinh, Viet.	E 2	35
Namib (des.), Namibia	J15	39
Nampa, Idaho	B 6	75
Nampula, Moz.	O15	39
Namsos, Nor.	C 2	24
Namur, Belg.	E 7	24
Nanaimo, B.C.	J 3	59
Nanakuli, Hawaii	D 2	74
Nanchang, China	H 6	37
Nancy, France	G 3	26
Nanda Devi (mt.), India	D 2	34
Nandi, Fiji	H 7	41
Nanga Parbat (mt.), India	D 1	34
Nanking, China	J 5	37
Nanning, China	G 7	37
Nan Shan (mts.), China	E 4	37
Nanterre, France	A 1	26
Nantes, France	C 4	26
Nanticoke, Pa.	K 3	101
Nantucket (isl.), Mass.	O 7	84
Napa, Calif.	C 5	69
Naperville, Ill.	E 2	76
Napier, N.Z.	M 6	40
Naples, Fla.	E 5	72
Naples, Italy	E 4	28
Napoleon, Ohio	B 3	98
Nara, Japan	E 5	36
Narayanganj, Bang.	G 4	34
Narberth, Pa.	M 6	101
Narbonne, France	E 6	26
Narmada (riv.), India	D 4	34
Narragansett, R.I.	J 7	84
Narragansett (bay), R.I.	J 6	84
N.A.S.A. Space Ctr., Tex.	K 2	105
Nasca, Peru	F 6	42
Nashua, N.H.	C 6	92
Nashville (cap.), Tenn.	H 2	104
Nassau (cap.), Bah.	C 1	45
Nässjö, Sweden	C 3	24
Natal, Braz.	O 5	42
Natal (prov.), S.Afr.	N17	39
Natchez, Miss.	B 7	87
Natchitoches, La.	D 3	81
Natick, Mass.	A 7	84
National City, Calif.	J11	69
Natural Bridges N.M., Utah	E 6	106
Naugatuck, Conn.	C 3	71
Nauru	G 6	41
Navajo N.M., Ariz.	E 2	67
Navarin (cape), U.S.S.R.	T 3	30
Navarre (reg.), Spain	F 1	27
Navojoa, Mex.	C 2	46
Nawabganj, Bang.	G 4	34
Náxos (isl.), Greece	D 4	29
Nayarit (state), Mex.	D 4	46
Nazaré, Braz.	N 6	42
Nazareth, Isr.	C 2	31
Ndalatando, Angola	K13	39
N'Djamena (cap.), Chad	K 9	38
Ndola, Zambia	M14	39
Neagh (lake), N.Ire.	C 3	23
Nebo (mt.), Jordan	D 4	31
Nebraska (state), U.S.		90
Nebraska City, Nebr.	J 4	90
Neckar (riv.), W.Ger.	C 4	25
Nederland, Tex.	K 8	105
Needham, Mass.	B 7	84
Needles, Calif.	L 9	69
Neenah, Wis.	J 7	111
Nefud (des.), Saudi Ar.	C-D 4	33
Negaunee, Mich.	B 2	85
Negev (reg.), Isr.	D 5	31
Negombo, Sri Lanka	D 7	34
Negro (riv.), S.A.	H 4	42
Negros (isl.), Phil.	G 4	36
Neisse (riv.), Europe	F 3	25
Neiva, Col.	F 3	42
Nejd (reg.), Saudi Ar.	C-E 4-5	33
Nelson, B.C.	J 5	59
Nelson (riv.), Man.	J 2	56
Nelson, N.Z.	L 6	40
Nenagh, Ire.	B 4	23
Neosho, Mo.	D 9	88
Nepal	E 3	34
Neptune, N.J.	E 3	93
Neptune City, N.J.	E 3	93
Ness (lake), Scot.	D 2	23
Netanya, Isr.	B 3	31
Netherlands		24
Netherlands Antilles	E 4, F 3	45
Neubrandenburg, E.Ger.	E 2	25
Neuchâtel, Switz.	B 2	27
Neuilly, France	B 1	26
Neumünster, W.Ger.	C 1	25
Neuquén, Arg.	G11	43
Neuse (riv.), N.C.	M 5	96
Neusiedler (lake), Europe	D 3	29
Neuss, W.Ger.	B 3	25
Neustadt, W.Ger.	B 4	25
Neu-Ulm, W.Ger.	D 4	25
Nevada, Mo.	D 7	88
Nevada (state), U.S.		91
Nevada, Sierra (mts.), Spain	E 4	27
Nevada, Sierra (mts.), U.S.	E-G 4-7	69
Nevers, France	E 4	26
Nevis (isl.), St.C.-N.-A.	F 3	45
New Albany, Ind.	F 8	77
New Albany, Miss.	G 2	87
New Amsterdam, Guyana	J 2	42
Newark, Calif.	K 3	69
Newark, Del.	L 2	83
Newark, Eng.	F 4	23
Newark, N.J.	B 2	93
Newark, N.Y.	G 4	95
Newark, Ohio	F 5	98
New Bedford, Mass.	K 6	84
Newberg, Oreg.	A 2	100
New Berlin, Wis.	K 2	111
New Bern, N.C.	L 4	96
Newberry, S.C.	D 3	102
New Braunfels, Tex.	F 8	105
New Brighton, Minn.	G 5	86
New Brighton, Pa.	B 4	101
New Britain, Conn.	E 2	71
New Britain (isl.), P.N.G.	F 6	41
New Brunswick (prov.), Can.		52
New Brunswick, N.J.	E 3	93
Newburgh, N.Y.	C 1	95
Newburyport, Mass.	L 1	84
New Caledonia	G 8	41
New Canaan, Conn.	B 4	71
New Carlisle, Ohio	C 6	98
New Carrollton, Md.	C 4	83
Newcastle, Aust.	J 6	40
Newcastle, Eng.	E 4	23
New Castle, Ind.	G 5	77
Newcastle, N.Br.	E 2	52
Newcastle, N.Ire.	D 3	23
New Castle, Pa.	B 3	101
Newcastle upon Tyne, Eng.	E 3	23
New City, N.Y.	C 2	95
New Cumberland, Pa.	J 5	101
New Delhi (cap.), India	D 3	34
New Fairfield, Conn.	B 3	71
Newfoundland (prov.), Can.		50
Newfoundland (isl.), Newf.	C 4	50
New Georgia (isl.), Sol. Is.	F 6	41
New Glasgow, N.S.	F 3	51
New Guinea (isl.), Pacific	D-E 6	41
New Hampshire (state), U.S.		92
New Hanover (isl.), P.N.G.	F 6	41
New Haven, Conn.	D 3	71
New Haven, Ind.	H 2	77
New Hebrides (Vanuatu)	G 7	41
New Hyde Park, N.Y.	A 2	95
New Iberia, La.	G 6	81
Newington, Conn.	E 2	71
New Ireland (isl.), P.N.G.	F 6	41
New Jersey (state), U.S.		93
New Kensington, Pa.	C 4	101
New London, Conn.	G 3	71
New London, Wis.		24
Newmarket, Ont.	E 3	55
New Martinsville, W.Va.	E 1	110
New Mexico (state), U.S.		94
New Milford, Conn.	B 2	71
New Milford, N.J.	B 1	93
Newnan, Ga.	C 4	73
New Orleans, La.	K 6	81
New Paltz, N.Y.	M 7	95
New Philadelphia, Ohio	G 5	98
New Plymouth, N.Z.	L 6	40
Newport, Ark.	H 2	68
Newport, Ky.	P 3	80
Newport, N.H.	B 5	92
Newport, Oreg.	C 3	100
Newport, R.I.	J 7	84
Newport, Tenn.	P 3	104
Newport, Vt.	C 1	107
Newport, Wales	E 5	23
Newport Beach, Calif.	D11	69
Newport News, Va.	L 6	108
New Port Richey, Fla.	D 3	72
New Providence (isl.), Bah.	C 1	45
New Providence, N.J.	E 2	93
New Roads, La.	G 5	81
New Rochelle, N.Y.	J 1	95
Newry, N.Ire.	C 3	23
New Shrewsbury, N.J.	E 3	93
New Siberian (isls.), U.S.S.R.	O-Q 2	30
New Smyrna Bch., Fla.	F 2	72
New South Wales (state), Aust.	G-J 6	40
Newton, Iowa	H 5	78
Newton, Kans.	E 3	79
Newton, Mass.	C 7	84
Newton, N.J.	D 1	93
Newton, N.C.	C 3	96
Newton Falls, Ohio	J 3	98
Newtown, Conn.	B 3	71
New Ulm, Minn.	D 6	86
New Waterford, N.S.	J 2	51
New Westminster, B.C.	K 3	59
New York, N.Y.	C 2	95
New York (state), U.S.		95
New Zealand	M 7	40
Nez Perce N.H.P., Idaho	A 3	75
Ngami (lake), Botswana	L16	39
N'Gaoundéré, Cameroon	J10	38
Niagara (riv.), N.A.	B 4	95
Niagara Falls, N.Y.	B 4	95
Niagara Falls, Ont.	E 4	55
Niamey (cap.), Niger	G 9	38
Nias (isl.), Indon.	B 5	36
Niassa (dist.), Moz.	O14	39
Nicaragua	C 2	46
Nice, France	G 6	26
Nicholasville, Ky.	H 5	80
Nicobar (isls.), India	G 7	34
Nicolet, Que.	E 3	53
Nicosia (cap.), Cyprus	B 2	33
Nicoya (gulf), C.R.	C 3	46
Nidwalden (canton), Switz.	D 2	27
Nieuw-Nickerie, Sur.	J 2	42
Niğde, Turkey	B 2	33
Niger	H 8	38
Niger (riv.), Afr.	G 9	38
Nigeria	H10	38
Nigríta, Greece	C 3	29
Niigata, Japan	E 3	36
Niihau (isl.), Hawaii	A 2	74
Nijmegen, Neth.	G 4	24
Nikolayev, U.S.S.R.	D 5	30
Nile (riv.), Afr.	N 7	38
Niles, Ill.	A 1	76
Niles, Mich.	C 7	85
Niles, Ohio	J 3	98
Nîmes, France	F 6	26

GLOSSARY OF ABBREVIATIONS

A

d. — Academy
B. — Air Force Base
1. — Afghanistan
— Africa
— Alabama
— Albania
— Algeria
1. — Alberta
er. — American
er. Samoa — American Samoa
— Antarctica
— Arabia
1. archipelago
— Argentina
2. — Arkansas
S. R. — Autonomous Soviet
    Socialist Republic
. — Austria
t. — Australia
— autonomous
Obl. — Autonomous Oblast
prov. — autonomous province
Reg. — Autonomous Region

B

1. — Bahamas
ng. — Bangladesh
b. — Barbados
. — British Columbia
1. — Beach
g. — Belgium
m. — Bermuda
— Bolivia
z. — Brazil
O.T. — British Indian Ocean Territory
g. — Bulgaria
.I. — British Virgin Islands

C

— cape
if. — California
. — Central America
nb. — Cambodia
. — Canada
. — capital
R. — Central African Republic
H. — Court House
in. — channel
an. Is. — Channel Islands
— county
— Colombia
o. — Colorado
n. — Connecticut
. — Costa Rica
— Center
Verde — Cape Verde
ch. — Czechoslovakia

D

. — Delaware
1. — Denmark
or. — depression
ot. — department
s. — desert
t. — district
m. Rep. — Dominican Republic

E

— East
ua. — Ecuador
Ger. — East Germany
Sal. — El Salvador
g. — England
Guin. — Equatorial Guinea
. — estuary
1. — Ethiopia

F

Falk. Is. — Falkland Islands
Fin. — Finland
Fla. — Florida
for. — forest
Fr. — French
Fr. Gui. — French Guiana
Fr. Poly. — French Polynesia
Ft. — Fort

G

Ga. — Georgia
Ger. — Germany
Greenl. — Greenland
Gt. — Great
Guad. — Guadeloupe
Guat. — Guatemala
Guin.-Biss. — Guinea-Bissau

H

hbr. — harbor
Hond. — Honduras
Hts. — Heights
Hung. — Hungary

I

i., isl. — island, isle
I.C. — Ivory Coast
Ice. — Iceland
Ill. — Illinois
Ind. — Indiana
Indon. — Indonesia
Int'l — International
Ire. — Ireland
is., isls. — islands
Isr. — Israel
isth. — isthmus

J

Jam. — Jamaica
Jct. — Junction

K

Kans. — Kansas
Ky. — Kentucky

L

La. — Louisiana
Lab. — Laboratory
Lak. — Lakeshore
Leb. — Lebanon
Lib. — Liberia
Liecht. — Liechtenstein
Lux. — Luxembourg

M

Mad. — Madagascar
Mart. — Martinique
Mass. — Massachusetts
M.C.A.S. — Marine Corps Air Station
Md. — Maryland
Mem. — Memorial
Mex. — Mexico
Mich. — Michigan
Minn. — Minnesota
Miss. — Mississippi
Mo. — Missouri
Mong. — Mongolia
Mont. — Montana
Mor. — Morocco
Moz. — Mozambique
mt., mtn. — mount, mountain
mts. — mountains

N

N. — North
N. A. — North America
N. B. — National Battlefield
N. B. P. — National Battlefield Park
N. Br. — New Brunswick
N. B. S. — National Battlefield Site
N. C. — North Carolina
N. Dak. — North Dakota
Nebr. — Nebraska
Neth. — Netherlands
Neth. Ant. — Netherlands Antilles
Nev. — Nevada
New Cal. — New Caledonia
Newf. — Newfoundland
New Hebr. — New Hebrides
N. H. — New Hampshire
N. H. P. — National Historical Park
N. H. S. — National Historic Site
Nic. — Nicaragua
Nig. — Nigeria
N. Ire. — Northern Ireland
N. J. — New Jersey
N. Korea — North Korea
N. Lab. — National Laboratory
N. Lak. — National Lakeshore
N. M. — National Monument
N. M. C. — Naval Missile Center
N. Mem. — National Memorial
N. Mem. Pk. — National Memorial Park
N. Mex. — New Mexico
N. M. P. — National Military Park
No. — Northern
Nor. — Norway
N. P. — National Park
N. R. A. — National Recreation Area
N. S. — Nova Scotia
N. Sea — National Seashore
N. W. T. — Northwest Territories
    (Canada)
N. Y. — New York
N. Z. — New Zealand

O

Obl. — Oblast
Okla. — Oklahoma
Ont. — Ontario
Oreg. — Oregon

P

Pa. — Pennsylvania
Pak. — Pakistan
Pan. — Panama
Par. — Paraguay
P. D. R. Y. — Peoples Democratic
    Republic of Yemen
P. E. I. — Prince Edward Island
pen. — peninsula
Phil. — Philippines
Pk. — Park
plat. — plateau
P. N. G. — Papua New Guinea
Pol. — Poland
Port. — Portuguese
P. R. — Puerto Rico
prom. — promontory
Prot. — Protectorate
prov. — province, provincial
Prov. Pk. — Provincial Park
Pt., Pte. — Point, Pointe

Q

Que. — Quebec

R

reg. — region
Rep. — Republic
res. — reservoir
Rhod. — Rhodesia
R. I. — Rhode Island
riv. — river
Rum. — Rumania

S

S. — South
S. A. — South America
S. Afr. — South Africa
São T. & P. — São Tomé & Príncipe
Sask. — Saskatchewan
Saudi Ar. — Saudi Arabia
S. C. — South Carolina
Scot. — Scotland
S. Dak. — South Dakota
Sea. — Seashore
Sen. — Senegal
Seych. — Seychelles
S. F. S. R. — Soviet Federated
    Socialist Republic
Sing. — Singapore
S. Korea — South Korea
S. Leone — Sierra Leone
Sol. Is. — Solomon Islands
Sp. — Spanish
Spr., Sprs. — Spring, Springs
S. S. R. — Soviet Socialist Republic
St., Ste. — Saint, Sainte
Sta. — Santa
St. C.-N.-A. — Saint Christopher-
    Nevis-Anguilla
Sto. — Santo
str. — strait
St. Vinc. & Grens. — St. Vincent &
    the Grenadines
Sur. — Suriname
S. W. Afr. — South-West Africa
Swaz. — Swaziland
Switz. — Switzerland

T

Tanz. — Tanzania
T. & C. Is. — Turks & Caicos Islands
Tenn. — Tennessee
terr. — territory
Tex. — Texas
Thai. — Thailand
T. & T. — Trinidad & Tobago
T. T. P. I. — Trust Territory of
    the Pacific Islands
Tun. — Tunisia

U

U. A. E. — United Arab Emirates
U. K. — United Kingdom
Urug. — Uruguay
U. S. — United States
U. S. S. R. — Union of Soviet
    Socialist Republics

V

Va. — Virginia
Ven. — Venezuela
V. I. — Virgin Islands (U.S.)
Viet. — Vietnam
Vill. — Village
vol. — volcano
Vt. — Vermont

W

W. — West
Wash. — Washington
W. Ger. — West Germany
W. I. — West Indies
Wis. — Wisconsin
W. Samoa — Western Samoa
W. Va. — West Virginia
Wyo. — Wyoming

Y

Y. A. R. — Yemen Arab Republic
Yugo. — Yugoslavia

Z

Zim. — Zimbabwe

POLITICAL DIVISION	GOVERNMENT	MAJOR PRODUCTS
AFGHANISTAN	A republic with a president, revolutionary council and cabinet; currently under Soviet military control.	Wheat, barley, corn, rice, sugar beets, nuts & seeds, fruits, cotton, tobacco; livestock; timber; natural gas, salt, copper, lead, talc, coal, lapis lazuli; hides & skins (karakul), wool, textiles, leather, carpets, cement.
ALBANIA	Soviet-type republic with a head of state, premier, cabinet and unicameral legislature; controlled by the Communist party.	Corn, tobacco, wheat, potatoes, cotton, sugar beets, fruits; livestock; fish; timber; petroleum, bitumen, lignite, nickel, copper, iron ore, chromite; textiles, wool, tobacco products, chemicals.
ALGERIA	Centralized republic under a president, premier, council of ministers, and an elected unicameral legislature.	Wheat, barley, oats, corn, grapes, olives, dates, figs, citrus fruits, vegetables, tobacco; fish; livestock; timber; iron ore, petroleum, phosphates, zinc, natural gas, mercury, lead; hides, wine, olive oil, cork, food & tobacco products, leather, textiles, chemicals, machinery, iron & steel, refined petroleum.
AMERICAN SAMOA	U.S. territory with an elected governor and bicameral legislature.	Taro, breadfruit, yams, bananas, arrowroot, pineapples, coconuts, oranges; fish; livestock; canned fish, copra, mats.
ANDORRA	Co-principality of the president of France and the Spanish bishop of Seo de Urgel, with an elected Syndic General and a general council.	Tobacco, potatoes, oats, barley; livestock; timber; iron ore, lead; dairy, tobacco, wood & wool products.
ANGOLA	Centralized republic under a president, assisted by a party central committee.	Coffee, corn, sugarcane, peanuts, tobacco, rice, palm products, cotton, sisal; iron ore, petroleum, diamonds; fish; livestock; timber; refined petroleum, cement, paper, tires, refined sugar, food products, chemicals.
ANTIGUA	Associated British state, with governor, prime minister, cabinet and bicameral legislature.	Sugar, cotton, rice, molasses, fruits, vegetables; fish; processed sugar and cotton, rum.
ARGENTINA	A republic with a president, at present under a military government.	Wheat, corn, millet, cotton, sugarcane, tobacco, fruits; livestock; timber; petroleum, natural gas, zinc, silver, lead, coal, iron ore, tungsten; wine, vegetable oils, dairy products, meat & meat products, wool, hides, textiles, wood and metal products, iron & steel, machinery, autos, chemicals, leather, petroleum products, cement.
AUSTRALIA	Independent British Commonwealth member with a governor-general, prime minister, cabinet, and a bicameral parliament, composed of a senate and a house of representatives.	Wheat, oats, barley, fruits, vegetables; livestock; gold, coal, petroleum, copper, iron, lead, silver, bauxite, uranium, zinc; timber, iron & steel, wool, electrical equipment, appliances, chemicals, petroleum products, optical & agricultural implements, machinery, textiles, leather, airplanes, engines, ships, processed meat, sugar, dairy products, building materials, autos, tires.
AUSTRIA	A federal republic with a president, chancellor, cabinet, and a partly elected bicameral parliament.	Rye, wheat, corn, oats, barley, potatoes, sugar beets, hops, flax, tobacco, grapes; livestock; timber; iron ore, copper, lead, graphite, coal, petroleum, salt, magnesite; wine, processed foods, dairy products, iron & steel, aluminum, machinery, tools, chemicals, paper, textiles, cement.
BAHAMAS	Independent British Commonwealth member, with a governor-general, prime minister, cabinet and bicameral general assembly.	Tomatoes, pineapples, sugarcane, vegetables, sponges, citrus fruits, bananas; fish, crawfish, shells; timber; salt; handcraft products, cement, pulpwood, processed fish, rum, refined petroleum, drugs.
BAHRAIN	Independent state with an emir, prime minister and cabinet.	Vegetables, fruits, dates; fish, shellfish; petroleum; refined petroleum, processed aluminum, electrical goods, cement, flour.
BANGLADESH	Independent republic in the British Commonwealth, with a president, prime minister, cabinet and unicameral parliament.	Rice, sugarcane, jute, cotton, oilseeds, tobacco, tea, chilies, fruit; timber; cattle, fish; natural gas, coal; textiles, hides & skins, flour, refined petroleum, steel, chemicals, refined sugar, handicrafts, paper, leather goods, jute products.

POLITICAL DIVISION	GOVERNMENT	MAJOR PRODUCTS
BARBADOS	Independent British Commonwealth member, with a governor-general, prime minister, cabinet and a bicameral parliament.	Sugarcane, vegetables, cotton; fish; manjak (asphalt); sugar, molasses, rum, edible oils, margarine.
BELGIUM	Constitutional, hereditary monarchy, with a king, premier, cabinet, and a bicameral parliament.	Wheat, rye, oats, barley, potatoes, sugar beets, tobacco, vegetables, fruit, hops; livestock, poultry; fish; coal, iron, zinc, lead, dolomite; coke, iron & steel, machinery, metal products, textiles, lace, glass, chemicals, petroleum & uranium refining, sugar, beer, paper, wine, wool, cut diamonds, dairy products, aircraft, cement, autos.
BELIZE	Internally self-governing British colony with governor, prime minister, cabinet and bicameral legislature.	Rice, corn, bananas, vegetables, citrus fruits, cocoa, sugarcane; cattle; hard and softwoods; fish, shellfish; rum, meat, fruit & fish products.
BENIN	Republic, with a head of state and a unicameral assembly.	Palm products, tobacco, peanuts, cotton, corn, copra, coffee, castor oil, kapok, millet; livestock; fish; gold, diamonds, bauxite, iron ore; oil seed milling, textiles.
BERMUDA	Partly self-governing British colony with a governor, prime minister, cabinet and a bicameral legislature.	Lily bulbs, onions, bananas, citrus fruits, vegetables, potatoes; coral; poultry, fish; limestone; perfume, pharmaceuticals, concrete.
BHUTAN	Monarchy with a king, council, and a unicameral assembly.	Rice, wheat, barley, millet, corn, fruits; timber; cattle, yaks; handicrafts, dairy products.
BOLIVIA	Centralized constitutional republic, presently ruled by a military junta.	Potatoes, corn, wheat, barley, rice, cassava, sugarcane, cotton, coffee, fruits; timber; livestock; tin, zinc, lead, silver, antimony, copper, natural gas, petroleum, tungsten, gold, sulphur; hides & skins, textiles, chemicals, cement, beer, tobacco products.
BOTSWANA	Constitutional republic within the British Commonwealth, with a president, cabinet, a unicameral parliament and an advisory house of chiefs.	Kaffir cotton, sorghum, millet, corn, wheat, beans, fruits & nuts; livestock; diamonds, nickel, copper, coal, salt, talc, manganese ore; hides & skins, meat & dairy products, leather goods, brewing.
BRAZIL	Federal republic with a president, vice-president, appointive cabinet and a bicameral legislature, at present ruled by decree.	Coffee, corn, rice, wheat, cotton, cocoa, sugarcane, soybeans, cassava, rubber, fibers, carnauba wax, medicinal plants, fruits & nuts, tobacco; livestock; timber; iron & manganese ore, diamonds, lead & zinc, bauxite, gold & silver, mica, asbestos, chromite, tungsten, petroleum, quartz, beryllium, copper, coal; meat products, hides, textiles, chemicals, petrochemicals, drugs, paper, lumber, machinery, autos, metal products, iron & steel, sugar, aluminum, tires, cement.
BRUNEI	Internally self-governing British protected sultanate, with a council of ministers.	Rice, sago, rubber, jelutong, cutch, tapioca, bananas; timber; livestock; petroleum, natural gas; boat building, cloth, brass and silverware, refined petroleum.
BULGARIA	Soviet-type republic with a cabinet, state council and unicameral parliament, which elects a presidium whose chairman is chief of state. Actual control is by the Communist party.	Wheat, corn, barley, cotton, tobacco, sugar beets, potatoes, seeds, fruits, vegetables; timber; livestock; fish; iron ore, copper, lead, coal, manganese, petroleum, zinc; food, leather & tobacco products, sugar, refined minerals & petroleum, textiles, wine, iron & steel, machinery, cement.
BURMA	One-party socialist republic with a unicameral assembly, prime minister and cabinet, and a state council with its chairman the president.	Rice, corn, cotton, pulses, sugarcane, tobacco, fruits & nuts, jute, rubber, sesame; livestock; timber (teak); petroleum, lead, zinc, tungsten, nickel, silver, copper, precious stones; sugar, food, tobacco & wood products, drugs, chemicals, textiles, cement, refined petroleum, steel.
BURUNDI	One-party republic with a president and revolutionary council.	Coffee, tea, cotton, corn, beans, fruits & nuts, sweet potatoes, sorghum; cattle; fish; timber; nickel; hides & skins, textiles, cement, beer, soap, shoes, food products.
CAMBODIA (KAMPUCHEA)	Communist state with a president, vice-president and a revolutionary council.	Rice, tobacco, corn, beans, sugarcane, rubber, cotton; cattle; fish; timber; phosphates, gold, precious stones; food products, textiles, sugar, glass, drugs.

POLITICAL DIVISION	GOVERNMENT	MAJOR PRODUCTS
CAMEROON	One-party republic, with a president, prime minister, cabinet, and unicameral elected assembly.	Cocoa, coffee, rubber, nuts, tea, rice, tobacco, palm products, cotton; livestock; fish; timber; bauxite, gold, petroleum; hides & skins, wood, rubber & tobacco products, textiles, beer, food products, palm oil.
CANADA	Independent confederation of the British Commonwealth, with a governor-general, prime minister, cabinet and a bicameral parliament, composed of an appointed senate and an elected house of commons.	Wheat, oats, barley, corn, potatoes, vegetables, sugar beets, tobacco, fruits, oilseeds; livestock, poultry; fish, shellfish; timber; furs; gold, copper, nickel, zinc, lead, silver, potash, molybdenum, platinum, iron ore, titanium, cobalt, radium, uranium, petroleum, natural gas, coal, asbestos, salt, gypsum, sulphur; hydroelectric power; foods, apparel, meat & dairy products, transportation equipment, iron & steel, aluminum, metal products, lumber, pulp, paper & wood products, textiles, electric goods, chemicals, autos, cement, processed minerals, refined petroleum, machinery.
CAPE VERDE	Republic, ruled by the president as chairman of the only political party; prime minister and cabinet.	Coffee, bananas, nuts, oilseeds, corn; livestock; salt, lime; hides & skins, preserved fish, sugar, cement.
CENTRAL AFRICAN REPUBLIC	Republic with a president, premier and cabinet.	Coffee, cotton, peanuts, tobacco, corn, rice, sorghum; timber; livestock; gold, diamonds, uranium; wood & palm products, textiles, flour, soap, beer.
CHAD	Republic with a head of state and a cabinet.	Millet, sorghum, rice, cotton, vegetables, dates, cassava, peanuts, gum arabic, ivory, ostrich feathers; livestock; fish; natron (salt); hides, cloth, meat products.
CHILE	Republic with a president as a member of a 4-man military junta, a cabinet and an advisory state council.	Cereal grains, seeds, sugar beets, potatoes, vegetables, fruits, tobacco; livestock; fish; timber; copper, nitrates, iron ore, manganese, silver, gold, molybdenum, zinc, coal, petroleum; chemicals, petrochemicals, wood & metal products, textiles, paper, pulp, drugs, wine, iron & steel, food & leather products, cement.
CHINA (PEOPLE'S REPUBLIC)	Nominal republic, ruled by a prime minister & cabinet (state council); controlled by Communist party's politburo, headed by the standing committee and its chairman.	Rice & cereal grains, soybeans, fruits, vegetables, nuts, oilseeds, tea, silk, cotton, sugarcane, tobacco; livestock, poultry; fish; timber; iron ore, petroleum, coal, tungsten, tin, antimony, magnetite, manganese, molybdenum, natural gas, mercury, bauxite, lead, zinc; meat & food products, textiles, apparel, ceramics, cement, iron & steel, machinery, metal products, aluminum, chemicals, vehicles, armaments.
CHINA (REPUBLIC OF): TAIWAN	Republic with a president, prime minister, cabinet, a legislative yuan and a national assembly, the latter electing the president.	Rice, sugarcane, tea, sweet potatoes, bananas, pineapples, mushrooms, soybeans, tobacco; livestock; fish; timber; coal, natural gas; food & wood products, cement, glass, chemicals, petrochemicals, steel, bicycles, sugar, electric & electronic goods, machinery, metal products, textiles, apparel.
COLOMBIA	A centralized federal republic with a president, vice-president, appointive cabinet, and elective bicameral congress.	Coffee, rice, cotton, sugarcane, bananas, cacao, wheat, corn, tobacco, rubber, fibers; livestock; fish; timber; petroleum, gold, platinum, emeralds, silver, salt; sugar, food & tobacco products, beer, textiles, cement, iron & steel, machinery, metal & leather products, chemicals, meat.
COMOROS	Republic with a president, premier, cabinet and unicameral legislature.	Sugarcane, vanilla, rice, root vegetables, copra, sisal, coffee, essential oils (ylang, citronella), cloves, cacao, perfume plants; timber; rum distilling.
CONGO	Republic of the French Community, with a president, premier and a national assembly.	Palm products, coffee, cocoa, bananas, tobacco, sugarcane, rice, corn, peanuts, fruits; livestock; timber; petroleum, potash, lead, zinc, gold; hardwoods & wood products, textiles, beer, cement, sugar, food products.
COOK ISLANDS	Internally self-governing state associated with New Zealand with a commissioner, prime minister, cabinet and legislative assembly.	Citrus fruits, coconuts, copra, oilseeds, tomatoes, arrowroot, pineapples, breadfruit, taro, kumaras, plantains, yams; mother-of-pearl, textiles, processed fruits.
COSTA RICA	Constitutional republic with president, cabinet and elected unicameral assembly.	Coffee, bananas, cocoa, corn, sugarcane, rice, potatoes, tobacco; cattle; tuna; timber; gold, salt, bauxite; dairy, tobacco & food products; electrical goods, beef, sugar, textiles, furniture, cement, apparel.

POLITICAL DIVISION	GOVERNMENT	MAJOR PRODUCTS
CUBA	Communist republic with a president, cabinet and an elected legislature, but with dictatorial powers held by the president and council of state.	Sugarcane, tobacco, coffee, rice, fruits; cattle; timber; fish; nickel, iron ore, chromite, manganese, copper; sugar, tobacco, meat & food products, textiles, cement, chemicals, steel, refined petroleum & metals, electrical goods, rum.
CYPRUS	British Commonwealth republic, at present divided into Greek and Turkish states, each with a president and unicameral legislature.	Wheat, barley, grapes, raisins, olives, potatoes, carobs, nuts, citrus fruits, tobacco, vegetables; fish; livestock; copper & concentrates, iron pyrites, asbestos, chromite, gypsum, marble; tobacco, leather & food products, cement, wine, textiles, refined petroleum.
CZECHOSLOVAKIA	Soviet-type republic with a president, premier, cabinet, bicameral legislature, and Czech and Slovak National Councils, with actual power residing in the Communist party presidium.	Wheat, rye, oats, barley, corn, sugar beets, potatoes; livestock; timber; coal, iron ore, magnesite, uranium, lead, salt; munitions, machinery, metal, rubber, leather & wood products, cement, iron & steel, textiles, shoes, porcelain, paper, chemicals, aircraft, autos, glass & glassware, beer, apparel, sugar, food products.
DENMARK	Constitutional, hereditary monarchy with a queen, a unicameral elective legislature and an appointed cabinet and premier.	Barley, oats, rye, wheat, potatoes, sugar beets, vegetables; poultry, livestock; stone, clay, iron ore; meat & meat products, dairy products, canned foods, beverages, machinery, transportation equipment, metal & rubber products, chemicals, apparel, shoes, furniture, glassware, earthenware, electrical goods, ships, cement, paper, tobacco products.
DJIBOUTI	Independent republic with a president, premier and a unicameral assembly.	Salt; hides & skins; livestock; boats.
DOMINICA	Independent British Commonwealth republic, with a president, prime minister, cabinet and a unicameral parliament.	Bananas, citrus fruits, timber, pumice, edible and essential oils, copra.
DOMINICAN REPUBLIC	Republic with a president, vice-president, appointed cabinet, and bicameral legislature.	Sugarcane, cacao, coffee, tobacco, bananas, rice, fruits, corn; cattle; lumber; nickel, bauxite; nickel & petroleum refining, chocolate, sugar, meat, cigars, textiles, cement, beer, flour, peanut oil, leather goods, rum.
ECUADOR	Constitutional republic with a president, cabinet and unicameral legislature.	Rice, cocoa, coffee, sugarcane, corn, bananas, cotton, cinchona; livestock; fish & shellfish; timber; petroleum, gold, silver; food, rubber, leather & wood products, textiles, toquilla (panama) hats, sugar, beer, cement, chemicals & petrochemicals, drugs, glass.
EGYPT	Arab republic with a president, appointed prime minister, cabinet, and a partly elected unicameral assembly.	Cotton, cereal grains, sugarcane, fruits, vegetables; livestock; fish; petroleum, phosphates, salt, iron ore, manganese, limestone; cotton ginning, iron & steel, refined petroleum, food processing, textiles, chemicals, cement, petrochemicals, sugar.
EL SALVADOR	Republic with a 5-man ruling junta, and a cabinet.	Coffee, cotton, cereal grains, cacao, tobacco, henequén, sugarcane; fish, shellfish; livestock; timber; silver; sugar, textiles, food products, drugs, chemicals, electric goods.
ENGLAND AND WALES	Integral part of the United Kingdom, with executive power nominally residing in the Crown, but actually exercised by the prime minister, cabinet and bicameral parliament, composed of a house of lords and a house of commons.	Potatoes, vegetables, cereal grains, hay, hops, fruits; livestock, poultry; fish; coal, petroleum, natural gas, iron ore, copper, lead, nickel, tin; dairy products, wool, cotton & linen textiles, electrical goods, vehicles, steel, scientific instruments, cutlery, foods & beverages, leather & tobacco products, apparel, chemicals, petrochemicals, pottery, china, machinery, locomotives, knitwear, drugs.
EQUATORIAL GUINEA	Nominal republic, with a president and a Supreme Military Council.	Cocoa, coffee, bananas, sugarcane, palm oil & kernels; timber, cabinet woods; fish; copra, beverages, soap.
ETHIOPIA	Military state, with the chairman of the military council as head of state, and a cabinet.	Coffee, wheat, corn, barley, durra, teff, pulses, oilseeds, chat, civet, fruits, vegetables, sugarcane, spices; poultry, livestock; gold, platinum; hides & skins, meat & food products, textiles, cement, sugar, refined petroleum, drugs.
FALKLAND ISLANDS	British colony with a governor, executive & legislative councils.	Oats, vegetables, hay; sheep; wool, hides & skins, tallow, animal & vegetable oil.
FIJI	Independent British Commonwealth member with a governor-general, prime minister, cabinet, and a bicameral parliament.	Sugarcane, coconuts, rice, fruits, cotton, rubber, ginger, oilseeds, vegetables, bananas, cocoa, corn, tobacco; livestock; timber; fish; gold, silver, manganese; sugar, copra, coconut oil, molasses, candlenut oil, cement, beer, meat products, flour, shipbuilding.

POLITICAL DIVISION	GOVERNMENT	MAJOR PRODUCTS
FINLAND	Constitutional republic with a president, premier, cabinet, and a unicameral parliament.	Hay, potatoes, cereal grains; livestock, poultry, reindeer; timber; fish; copper, iron ore, titanium, zinc, nickel; lumber, plywood, furniture, pulp, paper, wood products, textiles, food & dairy products, meat, chemicals, china, glass, machinery, ships, transportation equipment, electrical & metal products, vehicles, apparel, iron & steel.
FRANCE	A constitutional republic with a president, premier, bicameral elective legislature and appointive council of ministers.	Sugar beets, potatoes, cereal grains, turnips, fruits, nuts, grapes, buckwheat; livestock; fish; coal, iron ore, bauxite, pyrites, potash, salt, sulphur, natural gas; iron & steel, chemicals, machinery, metal & leather goods, autos, aircraft, ships, aluminum, porcelain, food & dairy products, apparel, cosmetics, perfumes, sugar, wines & spirits, electric & electronic goods, lace, silk, cotton, rayon, wool & linen textiles.
FRENCH GUIANA	Overseas department of France governed by a prefect with an elective general council.	Rice, bananas, sugarcane, corn, manioc; timber; livestock; shrimp; bauxite, gold; hides, shoes, rum, fish glue.
FRENCH POLYNESIA	Overseas territory of France, with a governor, government council, and an elected territorial assembly.	Coconuts, bananas, pineapples, oranges, vanilla, sugarcane, coffee, bamboo; fish; mother-of-pearl, sugar, rum, copra.
GABON	One-party republic of the French Community with a president, appointed prime minister, and unicameral national assembly.	Coffee, cocoa, rubber, corn, rice, bananas, cassava; timber; fish; manganese, uranium, petroleum, iron ore, gold, natural gas, lead, zinc, copper, diamonds, phosphates; refined petroleum, processed metals, textiles, plywood.
GAMBIA	Republic of the British Commonwealth, with a president, vice-president, cabinet and unicameral legislature.	Peanuts, rice, millet, sorghum, fruits, palm kernels; livestock; fish; textiles, peanut oil refining, fish processing, palm products, beverages.
GERMANY	Divided country with two governments. The western democratic Federal Republic has a president, chancellor, cabinet & bicameral parliament. The eastern Democratic Republic is ruled by the chairman of the state council, a prime minister & cabinet, & a unicameral legislature; actual power resides in the head of the Communist party.	Cereal grains, potatoes, sugar beets, fruits, hops; livestock; fish; timber; coal, lignite, iron ore, potash, salt, uranium, lead, zinc, natural gas, fluorspar; iron & steel, autos, bicycles, machinery, aluminum, cement, electrical & transportation equipment, ships, metal & electronic products, cotton & woolen textiles & yarn, rayon fiber, precision & optical instruments, shoes, apparel, food products, sugar, beer, wine, chemicals, sulphuric acid, soda, ammonia, synthetic rubber, drugs, petrochemicals.
GHANA	Republic of the British Commonwealth, with a president and a unicameral parliament.	Cocoa, coconuts, kola nuts, fruits, tobacco, coffee, peanuts, rubber; livestock; fish; timber; gold, diamonds, manganese, bauxite; aluminum, refined petroleum, textiles.
GIBRALTAR	Partly self-governing British colony, with governor, cabinet, house of assembly and local council.	Fish; ship repairing, beer, local food processing.
GREAT BRITAIN	See: England and Wales, Northern Ireland, Scotland.	
GREECE	Constitutional republic, with a president, premier, and unicameral parliament.	Cereal grains, tobacco, sugar beets, cotton, fruits, olives; livestock; sponges, fish; iron ore, emery, manganese, magnesite, marble, silver, nickel, bauxite, salt, chromite; textiles, olive oil, processed meat, fruit & vegetables, dairy, wood & leather products, steel, machinery, refined aluminum & petroleum, chemicals, wine, olive oil, cement, drugs.
GREENLAND (KALATDLIT-NUNAT)	Self-governing community of Denmark with a premier and elected legislature.	Grass for fodder; cod and other fish; sheep, furs; cryolite, lead, zinc; processed fish, skins.
GRENADA	Independent British Commonwealth member with a governor-general, premier and revolutionary council.	Cocoa, nutmeg, coffee, mace, limes, bananas, sugarcane, coconuts, vegetables, cotton; fish; livestock; timber; sugar, cotton ginning, copra, lime oil, rum, beer, cigarettes.
GUADELOUPE	Overseas department of France with a prefect and elected general council.	Sugarcane, bananas, pineapples, mangoes, avocados, coffee, cotton, sisal, cocoa, vanilla, cassava; fish; rum, sugar.
GUAM	Unincorporated U.S. territory, with an elected governor, advisory staff, and a unicameral legislature.	Coconuts, corn, bananas, citrus fruits, mangoes, papayas, breadfruit, sweet potatoes, cassava, vegetables, sugarcane, pineapples; livestock, poultry; fish; dairy & coconut products.

POLITICAL DIVISION	GOVERNMENT	MAJOR PRODUCTS
GUATEMALA	Republic with a president, cabinet and an elected unicameral congress.	Coffee, bananas, sugarcane, tobacco, rubber, cotton, chicle, abacá; fish; cattle; mahogany; nickel, zinc, lead; textiles, chemicals, essential oils, wood, metal & electric goods, processed meat & foods, sugar, hides & skins, apparel.
GUINEA	One party republic with a president, cabinet, premier and unicameral national assembly.	Rice, millet, coffee, kola nuts, peanuts, palm oil & kernels, quinine, pineapples, cassava, bananas; livestock; bauxite, iron ore, diamonds, gold; timber; hides & skins, textiles, wood & food products, cigarettes, aluminum.
GUINEA-BISSAU	Independent republic, with a state council under the president, and a one-party unicameral assembly.	Rice, palm kernels, palm oil, wax, peanuts, coconuts; hides and skins; fish; timber.
GUYANA	Republic within the British Commonwealth, with president, prime minister, cabinet, and unicameral assembly.	Sugarcane, corn, rice, coconuts, coffee, citrus & tropical fruits, cacao, balata, rubber; timber; livestock; shrimp; bauxite, diamonds, manganese, gemstones, gold; textiles, milled rice, beer, rum, lime oil, sugar, wood & pulp, molasses, aluminum.
HAITI	Nominal republic with president (for life), cabinet, and a unicameral legislature.	Coffee, sugarcane, sisal, cotton, fruits, rice, corn, cocoa; livestock; shellfish; bauxite; fiber, cement, essential oils, handicrafts, molasses, textiles, cement, sugar, soap, rum.
HONDURAS	Nominal republic, with a "provisional president."	Bananas, coffee, coconuts, tobacco, corn, beans, sugarcane, cotton, rice, henequén; mahogany; cattle; lead, zinc, gold, silver; meat & food products, sugar, lumber, vegetable oils.
HONG KONG	British colony ruled by a governor assisted by executive and legislative councils.	Rice, sugarcane, vegetables; fish; poultry, pigs; iron ore, wolfram, graphite; iron & steel, ships, enamel ware, apparel, textiles, cotton & plastic goods, toys, cameras, radios, electric & electronic goods.
HUNGARY	Soviet-type republic with a president, council, premier and unicameral assembly. Actual power is in the hands of the politburo of the Communist party.	Cereal grains, sugar beets, tobacco, grapes, fruits, potatoes; livestock, poultry; fish; timber; coal, petroleum, natural gas, iron ore, bauxite; flour, sugar, iron & steel, wines, textiles, chemicals, cotton & woolen goods, dairy, food, wood & paper products, machinery, tools & metal products, transportation equipment, drugs, aluminum, bicycles, cement.
ICELAND	A republic with a president, premier, an elective bicameral parliament, and an appointive cabinet.	Hay, potatoes, turnips, fruits, vegetables; livestock; fish; diatomite; dairy products, processed fish & fish products, meat, hides & skins, textiles, apparel, chemicals, cement, motors, vegetable oils.
INDIA	An independent republic within the British Commonwealth with a president, vice-president, prime minister, cabinet and a bicameral parliament.	Cereal grains, peanuts, seeds, tea, tobacco, opium, jute, cotton, rubber, coffee, sugarcane; fish; livestock; timber; coal, manganese, iron ore, petroleum, salt, mica, chromite, ilmenite, clay, copper, bauxite, gypsum; textiles, silk, cotton & jute fabrics, carpets, wood & metalwork, leather, cement, ships, refined petroleum, sugar, iron & steel, machinery, typewriters, aluminum, autos, transportation equipment, aircraft, chemicals.
INDONESIA	Republic with a president, cabinet and a legislature contained within a congress.	Rice, sugarcane, corn, coconuts, cassava, sweet potatoes, spices, tea, coffee, fruits, rubber, tobacco, cotton, kapok; livestock; fish; timber; tin, petroleum, iron ore, natural gas, salt, bauxite, nickel, copper; refined petroleum & products, sugar, cement, copra, textiles, paper, ships, chemicals, palm oil, food products, glass, rubber goods, autos.
IRAN	Republic with a president, prime minister and a unicameral legislature; ultimate power rests with a religious leader.	Cereal grains, cotton, dates, raisins, fruits, opium, sugar beets, nuts, tea, tobacco; livestock; fish; timber; petroleum, natural gas, copper, lead, coal, iron ore, salt; hides, wool, textiles, carpets, leather & tobacco products, caviar, sugar, glass, tools, vehicles, iron & steel, cement, aluminum, refined petroleum, metal products, chemicals & petrochemicals, vehicles, flour, processed foods.
IRAQ	Nominal republic headed by a president and a revolutionary council, and a legislature.	Dates, fruits, barley, wheat, rice, tobacco, cotton, vegetables, sorghum; livestock; petroleum, sulphur, salt; refined petroleum, cement, chemicals, drugs, hides & skins, wool, glass, textiles, processed foods, electrical equipment.
IRELAND	Republic with a president, prime minister, cabinet, and a partly-elected bicameral parliament.	Hay, potatoes, turnips, sugar beets, cereal grains; fish; livestock; lead, zinc, silver; tobacco, textiles, apparel, wood, clay, paper & metal products, machinery, dairy products, meat, processed foods, beer, malt, chemicals, vehicles.

POLITICAL DIVISION	GOVERNMENT	MAJOR PRODUCTS
ISRAEL	Republic with a president, prime minister, cabinet and elected unicameral parliament.	Wheat, cotton, tobacco, vegetables, fruits; livestock, poultry; fish; potash, salt, petroleum; textiles, apparel, processed foods, dairy products, glass, drugs, instruments, paper, metal, wood, rubber & leather products, polished diamonds, electric & electronic products, chemicals, wine, vehicles.
ITALY	Constitutional republic with a president, premier, a bicameral elective parliament and an appointive cabinet.	Cereal grains, sugar beets, potatoes, tomatoes, olives, grapes, citrus fruits, tobacco; timber; fish; livestock; natural gas, sulphur iron ore, coal, zinc, bauxite, mercury, marble; textiles, chemicals, wine, autos, machinery, electrical goods, sugar, olive oil, apparel, processed foods, petrochemicals, iron & steel, aluminum, shoes, transportation equipment.
IVORY COAST	One-party republic with a president, cabinet, and a unicameral legislature.	Coffee, cocoa, sugarcane, bananas, pineapples, nuts, rubber, cotton; tropical woods; livestock; fish; diamonds, iron ore; textiles, processed foods, lumber & wood products, refined petroleum, metal products, palm oil.
JAMAICA	Independent member of the British Commonwealth, with a governor-general, prime minister, cabinet, and bicameral parliament.	Sugarcane, bananas, tobacco, coconuts, coffee, citrus fruits, pimento, spices; fish; timber; bauxite, gypsum; rum, molasses, textiles, aluminum, copra, apparel, chemicals, processed foods, sugar, cement, metal, paper & rubber products.
JAPAN	Constitutional monarchy, with a prime minister, cabinet, and a bicameral diet. The duties of the emperor are merely ceremonial.	Rice, wheat, barley, potatoes, fruits, vegetables, sugarcane, hemp, tobacco, soybeans, tea; livestock; fish; timber; petroleum, iron ore, manganese, gold, silver, copper, coal, natural gas; textiles, silk, iron & steel, machinery, autos, ships, instruments, electric & electronic goods, paper, pulp, porcelain & earthenware, toys, sugar, chemicals, apparel, aluminum, fish products, metal products.
JORDAN	Constitutional monarchy, with a king, premier and assembly.	Wheat, barley, grapes, vegetables, fruits, olives; livestock; phosphates, potash, marble; wool, tobacco & leather products, cement, soap, olive oil, beverages, refined petroleum.
KENYA	One-party republic of the British Commonwealth, with a president, vice-president, cabinet, and unicameral national assembly.	Sisal, wheat, tea, coffee, pyrethrum, cotton, sugarcane, corn, peanuts, coconuts, wattle bark; livestock; timber; gold, silver, fluorspar, salt; sisal, meat & dairy products, sugar, cement, soda ash, hides & skins, petroleum products.
KIRIBATI	Republic of the British Commonwealth with a president, cabinet and a unicameral legislature.	Coconuts, breadfruit; phosphate of lime; pearl shell, fish; pigs, poultry; copra, palm products.
KOREA	Divided country with two governments. South Korea is a republic with a president, prime minister, cabinet & unicameral assembly. North Korea is ruled by the politburo of the Communist party, and has a president, prime minister & unicameral assembly.	Rice, barley, wheat, soybeans, tobacco, corn, cotton, fruits; timber; livestock; fish; tungsten, gold, silver, iron ore, copper, coal, petroleum, lead, graphite, kaolin; textiles, silk, apparel, electric & electronic goods, metal, rubber, paper, wood & petroleum products, chemicals, cement, machinery, iron & steel.
KUWAIT	Constitutional state with an emir, prime minister and cabinet, at present ruled by decree.	Pearls; fish; petroleum, natural gas; refined petroleum, ammonia, chemicals, fertilizer.
LAOS	Communist republic with a president, premier and appointed assembly, controlled by the party.	Rice, coffee, tea, citrus fruits, corn, cinchona, opium, potatoes, tobacco, cardamon, stick-lac; livestock; timber; tin; textiles, cigarettes, beverages, lumber, milled rice.
LEBANON	Republic with a president, an appointed premier and cabinet, and an elected unicameral assembly.	Wheat, barley, corn, potatoes, fruits, onions, vegetables, olives, tobacco; livestock; iron ore; textiles, metal & tobacco products, refined petroleum, chemicals, processed foods.
LESOTHO	Monarchy presently ruled by a prime minister (by decree), cabinet and assembly.	Cereal grains, beans, peas; livestock; diamonds; wool, mohair, hides & skins, carpets, textiles, shoes, candles, chemicals, jewelry, processed foods.
LIBERIA	Republic, presently ruled by a council and its chairman.	Rubber, rice, coffee, sugarcane, cocoa, palm oil & kernels, piassava; timber; fish, shrimp; iron ore, diamonds; petroleum products, cement, processed foods & rubber, lumber.
LIBYA	Arab republic ruled by a council with an appointed premier and cabinet.	Barley, wheat, olives, grapes, dates, vegetables, figs, peanuts, citrus fruits, almonds, esparto; livestock, sponge & tuna fishing; hides & skins; petroleum, natural gas; textiles, crude petroleum, processed foods, leather, olive oil.

POLITICAL DIVISION	GOVERNMENT	MAJOR PRODUCTS
LIECHTENSTEIN	Constitutional hereditary monarchy, with a prince, prime minister, and unicameral parliament.	Corn, wheat, potatoes, grapes; livestock; textiles, wine, leather, dairy products, ceramics, precision instruments, drugs, canned foods, postage stamps.
LUXEMBOURG	Constitutional monarchy with a grand duke, premier, cabinet, and a bicameral parliament.	Oats, potatoes, wheat, rye, grapes; livestock; timber; iron ore, slate, salt, gypsum; iron & steel, metal products, chemicals, tobacco, leather, wine, dairy products, rubber products, fertilizers, plastic goods.
MACAO	Partly autonomous Portuguese overseas province, under a governor, cabinet, and a legislative assembly.	Rice, vegetables; fish; cement, metal work, lumber, processed tobacco, matches, wine, textiles, fireworks.
MADAGASCAR	Republic of the French Community with a head of government, premier and legislature. Rule is by a military council.	Cassava, rice, corn, sweet potatoes, vanilla, cloves, sugarcane, coffee, bananas, beans, manioc, sisal, tobacco, raffia; timber; livestock; fish; graphite, mica, chromite; textiles, processed meat & foods, refined petroleum & petroleum products, cement, paper, sugar, beer, leather.
MALAWI	One-party republic of the British Commonwealth, with president (for life), cabinet, and unicameral assembly.	Tobacco, tea, cotton, sugarcane, tung nuts, pulses, sisal, corn, fruits, sorghum, rice, millet, peanuts, rubber; timber; livestock; bauxite, stone, gold; hides & skins, tung oil, meat, transportation equipment, machinery, ghee, sugar.
MALAYSIA	Constitutional monarchy of the British Commonwealth, with a paramount ruler, prime minister, cabinet and bicameral parliament.	Rubber, rice, coconuts, sugarcane, coffee, cocoa, pineapples, pepper, tea, tobacco, vegetables; livestock; fish; timber; tin, petroleum, copper, gold, antimony, bauxite, iron ore, manganese; rubber & wood products, steel, autos, refined petroleum, textiles, electric goods, sugar, fibers.
MALDIVES	Republic with a president and unicameral legislature.	Coconuts, corn, millet, pumpkins, sweet potatoes, fruits, nuts; fish, cowries; mats, boats, dried fish & fish products, handicrafts, copra, coir, ambergris, lace.
MALI	Republic ruled by a president and a military committee.	Millet, rice, sorghum, peanuts, corn, cotton, tobacco, nuts, sisal; livestock; fish; salt, gold, bauxite, iron ore, uranium; hides & skins, ceramics, jewelry, leather, rice mills, soap, processed fish & foods, textiles, sugar, cement, meat, fibers.
MALTA	An independent member of the British Commonwealth, with a president, prime minister, a cabinet and a unicameral parliament.	Wheat, barley, potatoes, onions, grapes, vegetables, fruits, cumin seed, cotton; livestock; fish; lace, wine, beer, cigarettes, buttons, pipes, gloves, textiles & yarn, flowers, ceramics, rubber & electronic goods, apparel.
MARTINIQUE	Overseas department of France, with a prefect and an elected general council.	Sugarcane, cocoa, mangoes, avocados, pineapples, bananas, coffee; fish; rum, sugar.
MAURITANIA	One-party republic, with a president-prime minister and governed by a military committee.	Cereal grains, beans, peanuts, melons, dates, gum arabic, henna, sweet potatoes; livestock; lobsters, fish; manganese, gypsum, iron ore, copper, salt; hides & skins, fish products.
MAURITIUS	Independent member of the British Commonwealth, with a governor-general, prime minister, cabinet, and unicameral parliament.	Sugarcane, aloe fiber, corn, coffee, vanilla beans, hemp, potatoes, sisal, peanuts, tea, yams, manioc, pineapples, tobacco, coconuts; molasses, rum, copra, sugar, dairy, tea & tobacco products, processed foods, textiles, fibers.
MAYOTTE	French territorial collectivity.	Vanilla, sisal, sugarcane, essential oils, rum; fish.
MEXICO	Constitutional federative republic with a president, council of ministers and a bicameral congress.	Grains, coffee, cotton, tomatoes, sugarcane, bananas, chicle, beans, oranges, henequén; timber; fish; shrimp; livestock; silver, gold, lead, zinc, petroleum, coal, sulphur, manganese, natural gas, iron ore, copper; sugar, hides, textiles, fibers, chemicals, aluminum, machinery, autos, refined petroleum, petrochemicals, cement, paper, drugs, metal products.
MONACO	Constitutional hereditary principality, with a prince and a unicameral council.	Principal revenue from gambling casino and tourism. Postage stamps, perfume, liqueurs, olive oil, oranges, chemicals, instruments, glass, processed foods, ceramics.
MONGOLIA	Soviet-type republic, with politburo chairman, council of ministers and a unicameral assembly.	Grains; livestock; coal, petroleum, lead, gold; dairy products, wool, hides & skins, processed foods, machinery, furs, meat & dairy products, textiles, leather, cement.

POLITICAL DIVISION	GOVERNMENT	MAJOR PRODUCTS
MOROCCO	Constitutional monarchy, with a king, an appointed prime minister, cabinet, and a unicameral parliament.	Wheat, barley, legumes, olives, nuts, citrus fruits, sugar beets, grapes, vegetables; cork, timber; livestock; fish; phosphates, iron ore, fluorite, coal, lead, zinc, manganese, petroleum, cobalt; textiles, carpets, pulp, wine, essential oils, olive oil, food & fish products, perfumes, wool.
MOZAMBIQUE	One-party republic with a president, cabinet and a unicameral assembly.	Sugarcane, cereal grains, coconuts, cotton, cashew nuts, peanuts, sisal, beans, tea, tobacco; timber; livestock; fish, shellfish; gold, coal, iron ore, bauxite; sugar, textiles, milled rice, cement, vegetable oils, processed foods & fish, copra.
NAMIBIA (SOUTH-WEST AFRICA)	South African controlled territory with an administrator-general.	Livestock; fish, shellfish; diamonds, copper, lead, zinc, salt, tin, manganese, vanadium, iron ore, cadmium, silver, fluorspar, tantalite, phosphate, sulfur, germanium; karakul wool & hides, fish processing, dairy products.
NAURU	Republic with a president, cabinet, and unicameral assembly.	Phosphates.
NEPAL	Constitutional monarchy, with king, prime minister, cabinet, and a unicameral parliament.	Rice, wheat, corn, millet, jute, sugarcane, potatoes, tea, oilseeds, medicinal herbs; timber; livestock; iron ore, copper; processed rice, tobacco, leather & wood products, textiles, sugar, chemicals, ghee, hides & skins.
NETHERLANDS	A constitutional, hereditary monarchy governed by the queen, a premier and cabinet, and a bicameral partly elected states general.	Potatoes, sugar beets, cereal grains, flax, legumes, flower bulbs, seeds, vegetables, fruits; livestock; fish; coal, petroleum, natural gas, salt; metal products, textiles, paper, chemicals, processed foods, apparel, ships, ceramics, cement, dairy, wood & tobacco products, petroleum products, machinery, electric & electronic products, transportation equipment, flowers, glass, processed diamonds.
NETHERLANDS ANTILLES	Self-governing part of Netherlands Union with governor, minister-president, cabinet & unicameral legislature (staten).	Fish; salt, phosphates; refined petroleum, petrochemicals, electronic equipment, textiles, beer.
NEW CALEDONIA	French overseas territory with a governor, government council & an elected territorial assembly.	Coconuts, coffee, cotton, corn, tobacco, bananas, pineapples, vegetables, rice; timber; livestock; nickel, chrome, manganese, iron ore, cobalt, copper, lead, silver, gold; canned meat, nickel & coffee processing, copra.
NEW ZEALAND	An independent member of the British Commonwealth governed by a governor-general, a prime minister, a cabinet and a unicameral parliament.	Cereal grains; livestock; timber; fish; gold, coal, mineral sands, limestone, petroleum, natural gas; meat, wool, hides & skins, apparel, timber & wood products, dairy products, food & tobacco products, autos, chemicals, fertilizers, beer, bricks, cement, electrical goods, machinery, paper, rubber & petroleum products.
NICARAGUA	Republic with a 5-person ruling junta and a state council.	Coffee, sugarcane, sesame, corn, bananas, rice, cocoa, tobacco, cotton, beans; cattle; fish; hardwoods; gold, copper, silver; sugar, wood products, meat products, textiles, cottonseed, chemicals, petroleum products, paper, food products.
NIGER	One-party republic, with a president, the head of a military government.	Millet, rice, manioc, peanuts, cotton, gum arabic, beans, sorghum; livestock; uranium, cassiterite, limesetone, salt, natron; hides & skins, meat, food & leather products, textiles, cement, peanut oil.
NIGERIA	Federal republic of the British Commonwealth, with a president, cabinet and a bicameral legislature.	Palm oil and kernels, cocoa, spices, tobacco, peanuts, cotton, rubber, soybeans, corn, rice, millet, coffee; livestock; fish, shrimp; timber; tin, coal, limestone, natural gas, petroleum, marble; metal products, cement, timber & wood products, textiles, beer, refined petroleum, hides & skins, processed foods & oils.
NIUE	Self-governing New Zealand dependency, with a prime minister and an assembly.	Limes, kumaras, passion fruit, bananas; copra, woven handicrafts.
NORTHERN IRELAND	Integral part of the United Kingdom with local government presently being reorganized.	Potatoes, oats, fruits, vegetables, barley, hay; poultry, livestock; limestone, basalt & igneous rocks, sand & gravel; linen, apparel, wool textiles, dairy products, meat & meat products, aircraft, machinery, tobacco, whiskey, electronic & transportation equipment, ships.

POLITICAL DIVISION	GOVERNMENT	MAJOR PRODUCTS
NORWAY	Constitutional hereditary monarchy, with a king, premier, cabinet, and unicamerally elected but bicamerally operating parliament.	Hay, oats, barley, wheat, rye, potatoes, fruits; livestock; fish; timber; iron ore, petroleum, nickel, zinc, natural gas, coal; pulp, paper, cellulose, ships, aluminum, machinery, chemicals, metal & electro-chemical products, transportation equipment, iron & steel, processed & canned fish & foods, textiles, wool, dairy products, leather, furs.
OMAN	An independent sultanate and an absolute monarchy, with an advisory cabinet.	Wheat, alfalfa, dates, limes, frankincense, coconuts, tobacco; livestock; fish; petroleum; dried fish & limes, ghee.
PACIFIC ISLANDS, TRUST TERR.	United States U.N. trusteeship, with a high commissioner, Northern Marianas is a U.S. commonwealth.	Vegetables, tropical fruits, coconuts; fish, trochus shell; poultry, livestock; copra, meat, handicrafts.
PAKISTAN	Federal republic with a president, presently ruled by a military council and an administrator.	Cereal grains, cotton, sugarcane, citrus fruits, dates, tobacco; livestock; fish; petroleum, salt, chromite, natural gas, gypsum, limestone; textiles, rugs, apparel, leather, wool, hides & skins, handicrafts, surgical instruments, sporting goods, sugar, chemicals, cement, iron & steel, refined petroleum, electric goods, tires.
PANAMA	Republic with a president and unicameral legislature.	Bananas, cocoa, abacá, coconuts, rice, sugarcane, coffee, fruits; fish, shrimp; livestock; timber; beer, sugar, wood & leather products, textiles, refined petroleum, processed foods, cement, apparel, drugs, fishmeal.
PAPUA NEW GUINEA	Independent British Commonwealth member, with a governor-general, prime minister, cabinet, and unicameral parliament.	Coconuts, coffee, copra, cocoa, rubber, sago, rice, kapok, sisal, bamboo, bananas; dairying, livestock, poultry, fish; timber; gold, silver, copper; tobacco products, boats, brewing.
PARAGUAY	Centralized republic with a president, an appointed cabinet and a bicameral congress.	Cotton, tobacco, sugarcane, cereal grains, yerba maté, soybeans, coffee, citrus fruits; livestock; timber, quebracho; beef, meat products, flour, refined petroleum products, oilcake & essential oils, hides, textiles, cement.
PERU	Republic, with a president, prime minister and a bicameral congress.	Cotton, sugarcane, potatoes, cereal grains, beans, potatoes, vegetables, fruits, coffee, guano; fish; livestock; petroleum, lead, zinc, copper, silver, gold, salt, iron ore; textiles, foodstuffs, fishmeal, sugar, cement, apparel, chemicals, refined metals, iron & steel, tires, hides & skins.
PHILIPPINES	Republic governed by a president and an assembly.	Rice, sugarcane, abacá, corn, tobacco, cocoa, coffee, nuts, kapok, peanuts, vegetables, maguey, rubber, fruits; livestock; fish; timber, gum resins; gold, iron ore, copper, chromite, silver, manganese, salt, coal, petroleum; sugar, textiles, rubber & tobacco products; lumber & wood products, autos, handicrafts, milled coconut oil & rice, fruit canning, copra, steel, cement, glass, chemicals, paper.
PITCAIRN ISLANDS	British colony, administered by the British high commissioner in New Zealand.	Fruits, vegetables; goats, poultry; handicrafts, postage stamps.
POLAND	Soviet-type republic with a chief of (council of) state, premier, & unicameral parliament; actual power lies with the politburo of the Communist party.	Potatoes, cereal grains, sugar beets; livestock; fish; timber; coal, lead, zinc, sulphur, iron ore, petroleum, copper, natural gas; iron & steel, chemicals & petrochemicals, coke, electric & electronic equipment, autos, ships, aluminum, metal, food & dairy products, sugar, glass, transportation equipment, cement, machinery, paper.
PORTUGAL	Constitutional republic with a president, premier, cabinet and unicameral parliament.	Cereal grains, potatoes, tomatoes, citrus fruits, grapes, olives; livestock; fish; timber; coal, wolfram, iron ore, sulphur, tungsten; wine, olive oil, cork, canned fish, food products, pulp, refined petroleum, ships, autos, textiles, electronic equipment, machinery, cement, steel.
PUERTO RICO	Self-governing commonwealth associated with the United States, with a governor, advisory council, and bicameral congress.	Sugarcane, tobacco, fruits, coconuts, coffee, cotton, vegetables; livestock; stone, sand & gravel; rum, molasses, sugar, canned fish & fruit, tobacco products, cement, leather, textiles, apparel, petrochemicals, metal & electronic products.
QATAR	Independent state with an emir and advisory council.	Dates, fruit, vegetables; shrimp, fish; livestock; natural gas, limestone, petroleum; fish products, cement, refined petroleum, petrochemicals.
RÉUNION	French overseas department, with a prefect and general council.	Sugarcane, tea, tobacco, vanilla, corn, manioc; livestock; essential oils, fruit & vegetable products, rum, sugar, molasses.

POLITICAL DIVISION	GOVERNMENT	MAJOR PRODUCTS
RUMANIA	A socialist republic, with a president, a state council, a cabinet, and a unicameral assembly; actual power resides in Communist party politburo.	Wheat, barley, corn, potatoes, sugar beets, tobacco, fruits; livestock; timber; petroleum, natural gas, coal, lignite, salt, iron ore, copper, bauxite, manganese, uranium; iron & steel, machinery, chemicals, lumber, wood & paper products, electric goods, refined petroleum, ships, cement, sugar, food products, textiles, metal products.
RWANDA	Nominal republic, at present under military rule by a president and advisory committee.	Coffee, cotton, rice, tea, corn, peanuts, pyrethrum, vegetables; livestock; cassiterite, tungsten, tantalite, beryl, wolfram; textiles, handicrafts, processed foods, beer, hides.
ST. CHRISTOPHER-NEVIS-ANGUILLA	Associated British state with a governor, prime minister, cabinet & unicameral assembly.	Sugarcane, cotton, rice, vegetables, tropical fruits, corn, yams, coconuts, livestock; fish, shellfish; salt; molasses.
ST. HELENA	British colony with a governor, legislative and executive councils.	Fruit, vegetables, lily bulbs, flax, sweet potatoes, potatoes; livestock, poultry; cordage, fibers, lace.
ST. LUCIA	Independent British Commonwealth state with a governor, prime minister, cabinet & unicameral assembly.	Bananas, coconuts, cocoa, tropical & citrus fruits, nutmeg, mace; fish; rum, copra, coconut oil, soap, cigarettes.
ST. PIERRE AND MIQUELON	French overseas department, with a prefect and general council.	Codfish; cattle; sienna earth, yellow ocher; fish products, furs.
ST. VINCENT AND THE GRENADINES	Independent state within the British Commonwealth with a governor-general, prime minister and a unicameral parliament.	Bananas, arrowroot, coconuts, rice, tropical fruits, cotton, corn, spices, peanuts, cocoa; fish; livestock; copra, rum, processed foods, cigarettes.
SAN MARINO	Republic with two regents, a cabinet, and unicameral council.	Wheat, fruits, grapes, vegetables; stone; livestock; textiles, postage stamps, wine, pottery, hides, cement, paper, leather.
SÃO TOMÉ AND PRÍNCIPE	One-party republic with a president, appointed premier and cabinet, and a unicameral assembly.	Cacao, coffee, coconuts, cinchona, bananas; livestock; palm oil, copra.
SAUDI ARABIA	Absolute monarchy under a king and advisory council of ministers; the king exercises all authority.	Dates, corn, wheat, coffee, fruits, henna, vegetables; fish; livestock; petroleum, gold, silver, gypsum, lead, copper; refined petroleum, petrochemicals, fertilizers, iron & steel, cement, meat & dairy products, hides, wool.
SCOTLAND	Integral part of United Kingdom, with secretary of state for Scotland in the U.K. cabinet, controlling local agriculture & fisheries, home & health, education, development, & economic planning.	Potatoes, sugar beets, wheat, barley, vegetables, fruits; livestock; fish, shellfish; petroleum, coal, iron ore, lead, stone; iron & steel, machinery, metal, dairy, tobacco & food products, textiles & yarn, watches, transportation equipment, electric & electronic goods, autos, ships, paper, whiskey, refined petroleum, aluminum, chemicals.
SENEGAL	One-party republic of the French Community, with a president, a prime minister, cabinet and unicameral assembly.	Millet, sorghum, rice, corn, peanuts, cotton, fruits, vegetables, sweet potatoes; livestock; fish; phosphates, titanium, limestone; textiles, processed fish & foods, cement, peanut oil & cakes, refined petroleum, chemicals.
SEYCHELLES	British Commonwealth republic with a president, appointed cabinet and unicameral assembly.	Coconuts, cinnamon, patchouli, vanilla, tea; fish, tortoise shell, guano; copra, coconut oil, dried fish, coir, essential oils.
SIERRA LEONE	One-party republic of the British Commonwealth, with a president, cabinet and unicameral parliament.	Palm oil & kernels, rice, coffee, kola nuts, ginger, vegetables, cassava, piassava, peanuts, cocoa; livestock; fish, shrimp; diamonds, iron ore, bauxite, rutile; palm products, rice & oil milling.
SINGAPORE	Republic of the British Commonwealth, with a president, prime minister, cabinet & unicameral parliament.	Rubber, coconuts, fruits, vegetables, rice, coffee, tapioca, tobacco; livestock; fish; tin, rubber & petroleum processing, rice & coconut milling, steel, chemicals, cement, lumber & wood products, textiles, bricks, palm & food products, paper, refined petroleum, drugs, ships, electric goods.
SOLOMON ISLANDS	Independent member of the British Commonwealth with a governor-general, prime minister, cabinet and a unicameral parliament.	Copra; livestock; fish; timber; copper, bauxite, nickel.
SOMALIA	One-party republic with a president and advisory cabinet, all power being held by the party's central committee.	Sugarcane, cotton, cereal grains, peanuts, sesame, tobacco, bananas, beans; livestock; fish, shellfish; salt; fish, food & meat products, sugar, textiles, hides & skins.

POLITICAL DIVISION	GOVERNMENT	MAJOR PRODUCTS
SOUTH AFRICA	Constitutional republic, with a state president, prime minister, cabinet & bicameral parliament. Transkei was granted independence in 1976; Bophuthatswana in 1977; Venda in 1979.	Cereal grains, tobacco, sugarcane; fruits, peanuts; livestock; fish, lobsters; gold, coal, diamonds, copper, asbestos, manganese, limestone, platinum, chromite, iron ore, vanadium, tin, antimony, uranium; timber; chemicals, wool, iron & steel, machinery, apparel, textiles, fish & food products, sugar, aluminum, metal products, hides, autos, cement, transportation equipment, dairy products.
SPAIN	Monarchy with a king, premier and cabinet, and a bicameral parliament.	Cereal grains, potatoes, legumes, citrus fruits, vegetables, olives, grapes, sugar beets, esparto, flax, hemp, pulses, nuts, sugarcane; livestock, poultry; fish; timber; coal, lignite, iron ore & pyrites, lead, zinc, mercury, copper, uranium, gypsum; textiles, paper, cement, hides, wine, olive oil, processed foods, cork, machinery, chemicals, leather, autos, refined petroleum, apparel, silk, shoes, processed foods & fruit, iron & steel.
SRI LANKA (CEYLON)	Independent republic of the British Commonwealth, with a president, a prime minister, a cabinet and a unicameral assembly.	Tea, coconuts, rubber, rice, cotton, spices, cocoa, nuts, sugarcane, fruits; fish; livestock; graphite, mineral sands, ilmenite, gem stones, limestone, salt, pearls; copra, plywood, leather, shoes, glass, steel, acetic acid, ceramics, quinine, strychnine, chemicals, drugs, textiles, cement, beer, refined petroleum, coconut & tobacco products, paper, apparel.
SUDAN	Republic with a president, cabinet and unicameral assembly. Local autonomy has been granted the southern provinces.	Cotton, cereal grains, gum arabic, oilseeds, senna, castor beans, resins, peanuts, sesame, dom & shea nuts, dates; livestock; ivory, trochus shell, mother-of-pearl; iron ore, manganese, chromite, salt, gold; textiles, cement, hides & skins, cottonseed, oilcake, sugar, leather, paint, soap.
SURINAME	Independent republic with a president, premier, cabinet, and elective unicameral parliament.	Rice, citrus fruits, coconuts, coffee, bananas, sugarcane, cacao, balata, corn, tobacco; livestock; shrimp; timber, balata; gold, bauxite; sugar, rum, lumber & plywood, molasses, aluminum, food & dairy products.
SWAZILAND	British Commonwealth monarchy, with a king, prime minister, cabinet and bicameral parliament.	Tobacco, corn, peanuts, sugarcane, sorghum, cotton, rice, pineapples, citrus fruits; livestock; timber; asbestos, iron ore, coal; meat & dairy products, sugar, pulp, canned fruits, textiles, hides & skins, wood & tobacco products.
SWEDEN	A constitutional hereditary monarchy with a titular king, prime minister, cabinet and a unicameral parliament.	Hay, sugar beets, potatoes, oilseeds, oats, wheat, rye, barley; timber; livestock; fish; iron ore, zinc, copper, lead; lumber, paper & wood products, machinery, textiles, iron & steel, metal & electric goods, chemicals, dairy, food & tobacco products, porcelain, glass, ships, furs, transportation equipment, matches, autos, munitions, liquor, instruments.
SWITZERLAND	Federal republic with a president, vice-president & executive federal council, & a bicameral elected federal assembly.	Cereal grains, sugar beets, potatoes, vegetables, fruits, tobacco; livestock; timber; salt, iron ore, manganese; dairy & tobacco products, watches & clocks, electric & glass products, instruments, jewelry, machinery, metal products, chocolate, wine, drugs, textiles & yarn, chemicals, aluminum, iron & steel, cement, sugar, meat, apparel, dyes, foods.
SYRIA	Arab republic with a president, premier, and unicameral legislative people's council.	Cereal grains, cotton, vegetables, olives, grapes, sugar beets, tobacco; livestock; petroleum, natural gas, phosphates, gypsum; leather, textiles, cement, refined petroleum, wool, hides & skins, sugar, processed foods & oils, apparel, glass, tobacco goods.
TANZANIA	One-party united republic of the British Commonwealth, with a president, vice-president, prime minister, cabinet and unicameral parlament proportionately representing Tanganyika and Zanzibar.	Sisal, fruits, cocoa, coconuts, cotton, cloves, pyrethrum, spices, coffee, tobacco, nuts, tea, oilseeds, sugarcane; livestock; hides & skins; diamonds, gold, phosphates, mica, salt, tin, gem stones; processed foods, cement, textiles, refined petroleum, copra, hides & skins, sugar, dairy & wood products, cordage, rolled iron & aluminum.
THAILAND (SIAM)	Constitutional monarchy, at present under a prime minister and a bicameral assembly.	Rice, coconuts, sugarcane, rubber, peanuts, tobacco, tapioca, jute, kenaf, cotton, corn; teak & other timber; livestock; fish; tin, wolfram, lead; lac, sugar, cement, textiles, tobacco & petroleum products, paper.
TOGO	One-party republic with a president and a unicameral assembly.	Palm oil & kernels, manioc, kapok, cocoa, coconuts, yams, cereal grains, coffee, cotton, peanuts, nuts, cassava; livestock; timber; phosphates, limestone; textiles, copra, cement.

POLITICAL DIVISION	GOVERNMENT	MAJOR PRODUCTS
TOKELAU	An island territory of New Zealand governed by an administrator.	Coconuts, fiber, taro; pigs, chickens; fish; hats, mats, copra.
TONGA	Constitutional British Commonwealth monarchy, with a king, appointed prime minister, and unicameral assembly.	Coconuts, bananas, yams, breadfruit, taro, cassava, papayas, pineapples, melons, tobacco, corn, peanuts, candlenuts; fish; livestock, poultry; copra, processed fruits.
TRINIDAD AND TOBAGO	Independent British Commonwealth republic with a president, prime minister, cabinet and bicameral parliament.	Coffee, cocoa, coconuts, sugarcane, citrus fruits; cattle; timber; petroleum, natural gas, asphalt, coal, clay; rum, textiles, sugar, chemicals, plastic, glass, clay, wood & food products, cement, electric goods, refined petroleum.
TUNISIA	Republic with a president (for life), an appointed premier and cabinet, and an elective unicameral assembly.	Cereal grains, grapes, esparto, olives, vegetables, nuts, fruits, dates; cork, timber; livestock; fish; phosphates, petroleum, iron ore, lead, zinc; flour, wine, olive oil, sugar, wool, pottery, leather, textiles, food processing, chemicals, iron & steel, paper, refined petroleum, metal & electric goods.
TURKEY	Nominal republic, at present governed by a six-man council.	Tobacco, cereal grains, cotton, fruits, opium, seeds, olives, nuts, sugar beets; livestock; fish; timber; chromite, iron ore, petroleum, copper, coal, lignite; textiles, iron & steel, chemicals, refined petroleum, rugs, paper, olive oil, wool, furs, sugar, mohair, silk, cement, skins.
TUVALU	Independent member of the British Commonwealth with a governor-general, prime minister and a unicameral parliament.	Copra, fish, handicrafts, postage stamps.
UGANDA	Republic of the British Commonwealth at present governed by a six-man military commission.	Cotton, coffee, tea, plaintains, sisal, peanuts, millet, corn, tobacco, sugarcane; livestock; salt, copper, gold, phosphates, tin; cement, beverages, sugar, chemicals, smelted copper, processed foods, textiles, hides & skins, steel.
U.S.S.R.	Federation of 15 union republics with a bicameral Supreme Soviet, which elects the presidium & council of ministers. Real power is largely exercised by the politburo & secretariat (under its general secretary) of the central committee of the Communist party.	Cereal grains, sugar beets, cotton, flax, potatoes, seeds, vegetables, tobacco; livestock; fish; timber; petroleum, natural gas, bauxite, uranium, platinum, iron ore, lead, zinc, copper, phosphates, mercury, gold, manganese, nickel, chromite, asbestos, potash; iron & steel, machinery, chemicals, refined petroleum, petrochemicals, ships, autos, aircraft, lumber & wood products, meat & dairy products, textiles, wool, sugar, tools & metal products, aluminum, furs, cement, paper, electric goods, instruments, transportation equipment, foods & beverages.
UNITED ARAB EMIRATES	Constitutional Arab federation of seven sheikhdoms, with a president, vice-president, premier and cabinet and a unicameral assembly.	Dates, cereal grains, vegetables; livestock; fish, pearl fishing; petroleum; cement, refined petroleum, petrochemicals, postage stamps, dried fish.
UNITED KINGDOM	See: England and Wales, Northern Ireland, Scotland.	
UNITED STATES	Federal republic with a president, vice-president, an appointed cabinet, and a bicameral congress (senate and house of representatives). It consists of 50 states, each with a governor and a state legislature (all except Nebraska being bicameral).	Cereal grains, hay, soybeans, potatoes, peanuts, sugar beets, sugarcane, vegetables, nuts, fruits, cotton, tobacco, flax; livestock, poultry; fish, shellfish; timber; petroleum, natural gas, coal, iron ore, copper, lead, zinc, gold, silver, molybdenum, bauxite, gypsum, phosphates, sulphur, stone, sand & gravel; iron & steel, machinery, transportation equipment, metal products, electric & electronic goods, autos, ships, aircraft, munitions, chemicals, tobacco, leather, rubber & plastic products, glass, wool, textiles, cement, food & dairy products, lumber & wood products, paper, refined petroleum, petrochemicals.
UPPER VOLTA	Republic with a president, premier and a unicameral assembly.	Cereal grains, sweet potatoes, peanuts, cassava, karite (shea nuts), vegetables, cotton, sisal, sesame, tea; livestock; gold, manganese, copper; hides & skins, meat products, sugar, flour, textiles, processed foods & oils, soap, cigarettes.
URUGUAY	A republic governed by a president, cabinet and a council of state.	Cereal grains, seeds, peanuts, fruits, hops, sugar beets, grapes, tobacco; livestock, meat & meat products, hides, wool, textiles, leather, wines, chemicals, refined petroleum, aluminum, steel, cement, sugar, metal products.

POLITICAL DIVISION	GOVERNMENT	MAJOR PRODUCTS
VANUATU	Republic with a president, premier, cabinet and unicameral assembly.	Coconuts, cocoa, coffee, bananas, yams, taro, manioc, fruits; timber; cattle; fish, trochus shell; manganese; meat and fish products, copra, lumber.
VATICAN CITY	The Pope exercises absolute legislative, executive & judicial power.	Postage stamps, religious articles.
VENEZUELA	Constitutional federal republic with a president, appointive cabinet, and an elected bicameral congress.	Coffee, cotton, cocoa, sugarcane, cereal grains, tobacco, beans, sisal, balata, rubber, bananas; livestock; fish, shrimp; petroleum, natural gas, iron ore, gold, coal, phosphates, nickel, salt, diamonds; leather, rubber, metal & wood products, sugar, food, dairy & meat products, vehicles, chemicals, refined petroleum, petrochemicals, paper, steel, transportation equipment, apparel.
VIETNAM	Communist republic with a president, premier and unicameral assembly; actual rule is by the party's central committee and politburo.	Rice, corn, sugarcane, coffee, fruits, nuts, vegetables, tea, manioc, peanuts, sweet potatoes, tobacco, cotton, rubber, silk; livestock, poultry; fish, shellfish; timber; coal, iron ore, chromite, uranium, phosphates, gold, tin; paper, textiles, chemicals, machinery, tobacco, lumber & wood products, sugar, processed foods, glass, beer, handicrafts, steel.
VIRGIN ISLANDS (BR.)	British colony with an administrator, chief minister and councils.	Bananas, tropical fruits, coconuts, vegetables; livestock, poultry; fish, turtles; handicrafts, rum, petroleum refining.
VIRGIN ISLANDS (U.S.)	Unincorporated U.S. territory with an elected governor & unicameral legislature.	Vegetables, sugarcane, citrus fruits, coconuts; cattle; fish; rum, bay rum & oil, molasses, handicrafts, sugar, lime juice, hides, bitters.
WALLIS & FUTUNA	French overseas territory with an administrator, & a local council and assembly.	Coconuts, bananas, taro, yams, cassava, arrowroot, vegetables; livestock, poultry; fish, trochus shell; copra, handicrafts.
WESTERN SAMOA	Independent member of the British Commonwealth, with a head of state, prime minister, cabinet and unicameral legislative assembly.	Breadfruit, coconuts, coffee, fruits, seeds, yams, pawpaws, cocoa, bananas, taro; fish; timber; livestock; copra, handicrafts, hides, lumber, processed foods, apparel, beverages, soap.
YEMEN ARAB REP.	Arab republic with a president, premier and cabinet, and an advisory military council.	Coffee, cereal grains, cotton, grapes, fruits, qat, sesame; cattle; fish; rock salt; textiles, hides, leather, handicrafts.
YEMEN, PEOPLES DEM. REP. OF	One-party republic with a head of state, prime minister, cabinet and a presidium appointed by a unicameral parliament.	Dates, cereal grains; coffee, qat, gums, tobacco, cotton, fruit, sesame; livestock; fish; salt; ship bunkering, refined petroleum, hides & skins, textiles, fish products.
YUGOSLAVIA	A socialist federal republic with a president, premier and federal executive council, and a bicameral assembly. Actually ruled by the Communist party.	Cereal grains, sugar beets, tobacco, potatoes, seeds, hemp, nuts, fruits; livestock; fish; timber; coal, gold, iron ore, manganese, petroleum, bauxite, chromite, mercury, antimony, copper, lead, zinc, salt; textiles, lumber & wood products, cement, sugar, food & metal products, machinery, chemicals, iron & steel, ships, wine.
ZAIRE	One-party republic with a president, premier, executive council and unicameral legislative council; rule is by decree.	Palm oil & kernels, cotton, coffee, tea, cocoa, rice, sugarcane, rubber; livestock; ivory; timber; copper, diamonds, gold, cobalt, tantalite, petroleum, zinc, manganese, bauxite, cassiterite; textiles, processed foods, sugar, rubber products.
ZAMBIA	One-party republic of the British Commonwealth, with a president, prime minister, cabinet, and a unicameral assembly.	Cereal grains, tobacco, peanuts, cassava, sugarcane, fruits, cotton; timber; fish; livestock; copper, lead, coal, manganese, zinc, cobalt; iron & steel, metal & tobacco products, textiles, chemicals, refined petroleum & copper, processed foods & beverages, sugar, drugs, tires.
ZIMBABWE	A republic with a president, prime minister, cabinet and a bicameral parliament.	Corn, tobacco, peanuts, wheat, cotton, tea, sugarcane, citrus fruits; livestock; fish; copper, gold, asbestos, chromite, coal; textiles, apparel, cigarettes, wood, food, dairy & rubber products, meat & meat products, sugar, iron & steel, vehicles, electrical goods, metal products, chemicals, hides.

SELF-SUFFICIENCY IN RAW MATERIALS

	IRON	BAUXITE (ALUMINUM ORE)	ZINC	COPPER	LEAD	MANGANESE	CHROMIUM	MERCURY	PLATINUM	TUNGSTEN	TIN	NICKEL	MAGNESITE	SULFUR	PHOSPHATES	POTASH	COAL
United States																	
Canada																	
United Kingdom																	
France																	
Germany																	
U.S.S.R.																	
India																	
Japan																	

	PETROLEUM	RUBBER	COFFEE	SUGAR	WHEAT	CORN	RICE	MEAT	FISH	DAIRY PROD.	TOBACCO	COTTON	WOOL	SILK	FOREST PROD.	CHEMICALS
United States																
Canada																
United Kingdom																
France																
Germany																
U.S.S.R.																
India																
Japan																

KEY: GREEN AREAS INDICATE DEGREE OF SELF-SUFFICIENCY

= SURPLUS SUPPLY

Prepared by C.S. HAMMOND & Co. Inc., N.Y.

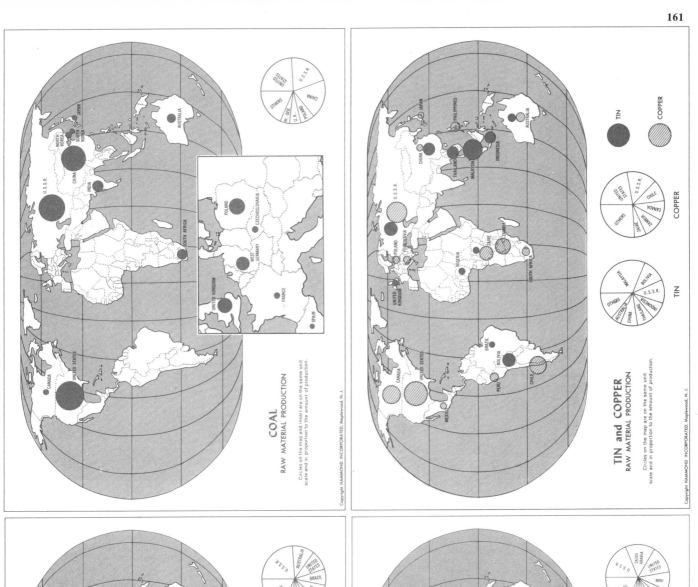

COAL
RAW MATERIAL PRODUCTION

Circles on the map and insert are on the same unit scale and in proportion to the amount of production.

Copyright HAMMOND INCORPORATED, Maplewood, N. J.

TIN and COPPER
RAW MATERIAL PRODUCTION

Circles on the map are on the same unit scale and in proportion to the amount of production.

Copyright HAMMOND INCORPORATED, Maplewood, N. J.

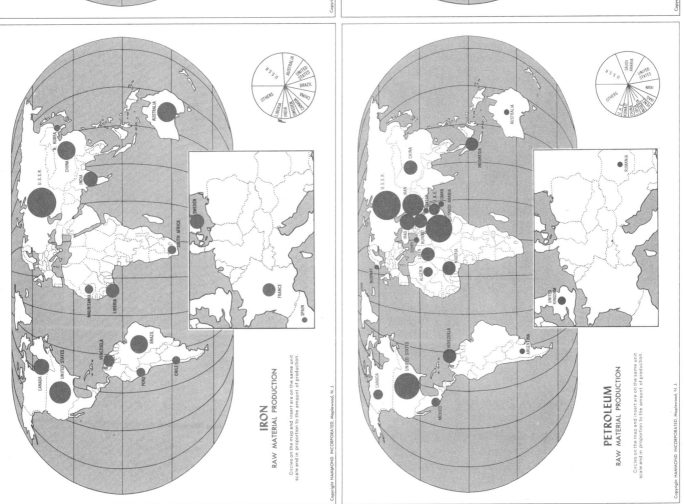

IRON
RAW MATERIAL PRODUCTION

Circles on the map and insert are on the same unit scale and in proportion to the amount of production.

Copyright HAMMOND INCORPORATED, Maplewood, N. J.

PETROLEUM
RAW MATERIAL PRODUCTION

Circles on the map and insert are on the same unit scale and in proportion to the amount of production.

Copyright HAMMOND INCORPORATED, Maplewood, N. J.

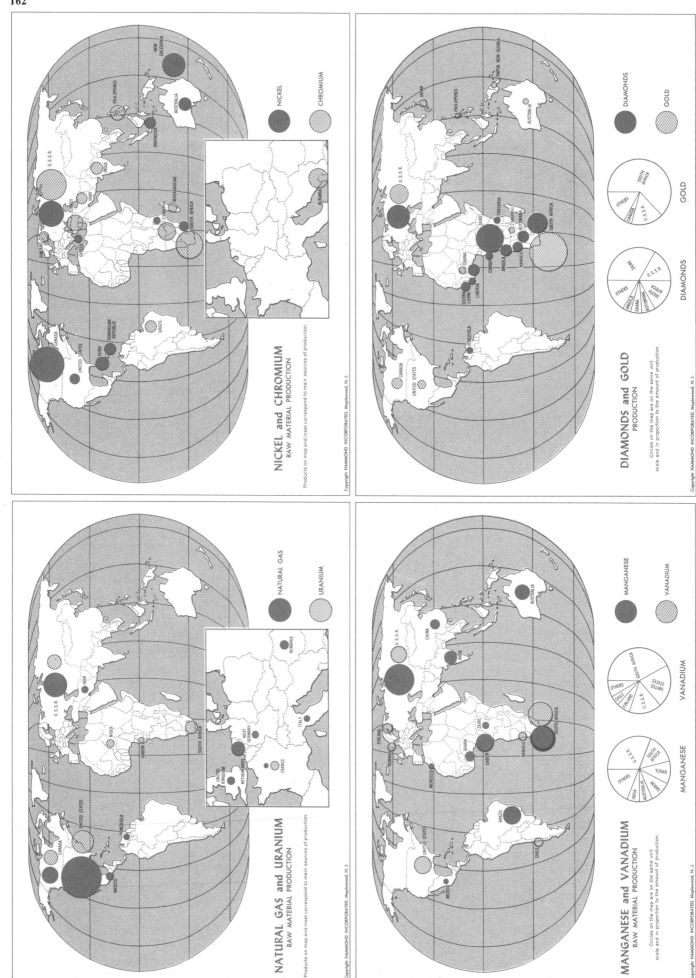

162

NICKEL and CHROMIUM
RAW MATERIAL PRODUCTION

Products on map and inset correspond to main sources of production.

NICKEL

CHROMIUM

Copyright HAMMOND INCORPORATED, Maplewood, N. J.

DIAMONDS and GOLD
PRODUCTION

DIAMONDS

GOLD

GOLD

DIAMONDS

Circles on the map are on the same unit
scale and in proportion to the amount of production.

Copyright HAMMOND INCORPORATED, Maplewood, N. J.

NATURAL GAS and URANIUM
RAW MATERIAL PRODUCTION

Products on map and inset correspond to main sources of production.

NATURAL GAS

URANIUM

Copyright HAMMOND INCORPORATED, Maplewood, N. J.

MANGANESE and VANADIUM
RAW MATERIAL PRODUCTION

MANGANESE

VANADIUM

VANADIUM

MANGANESE

Circles on the map are on the same unit
scale and in proportion to the amount of production.

Copyright HAMMOND INCORPORATED, Maplewood, N. J.

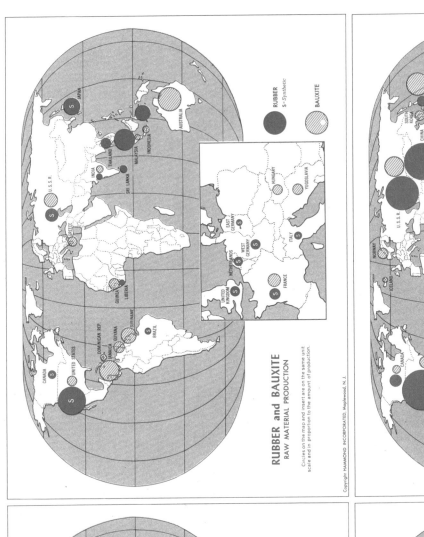

LEAD and ZINC
RAW MATERIAL PRODUCTION

LEAD
ZINC

Circles on the map and insert are on the same unit scale and in proportion to the amount of production.

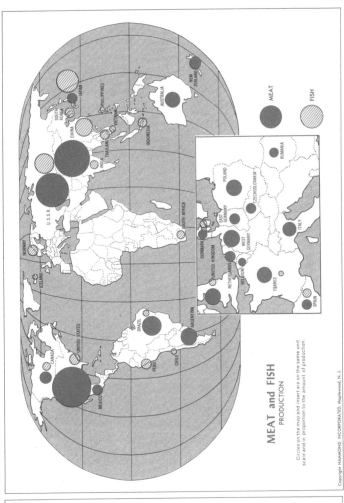

RUBBER and BAUXITE
RAW MATERIAL PRODUCTION

RUBBER
S—Synthetic
BAUXITE

Circles on the map and insert are on the same unit scale and in proportion to the amount of production.

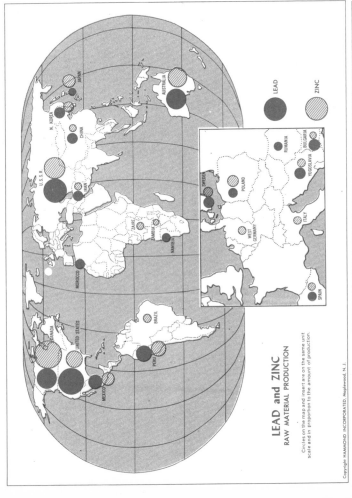

WHEAT
PRODUCTION

Circles on the map and insert are on the same unit scale and in proportion to the amount of production.

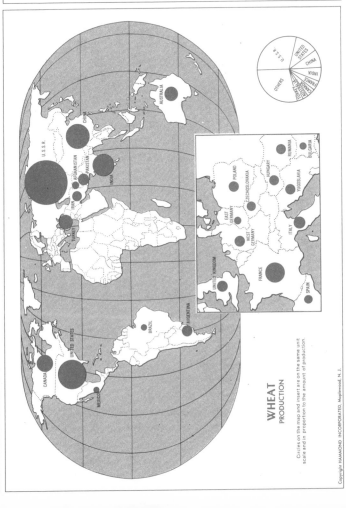

MEAT and FISH
PRODUCTION

MEAT
FISH

Circles on the map and insert are on the same unit scale and in proportion to the amount of production.

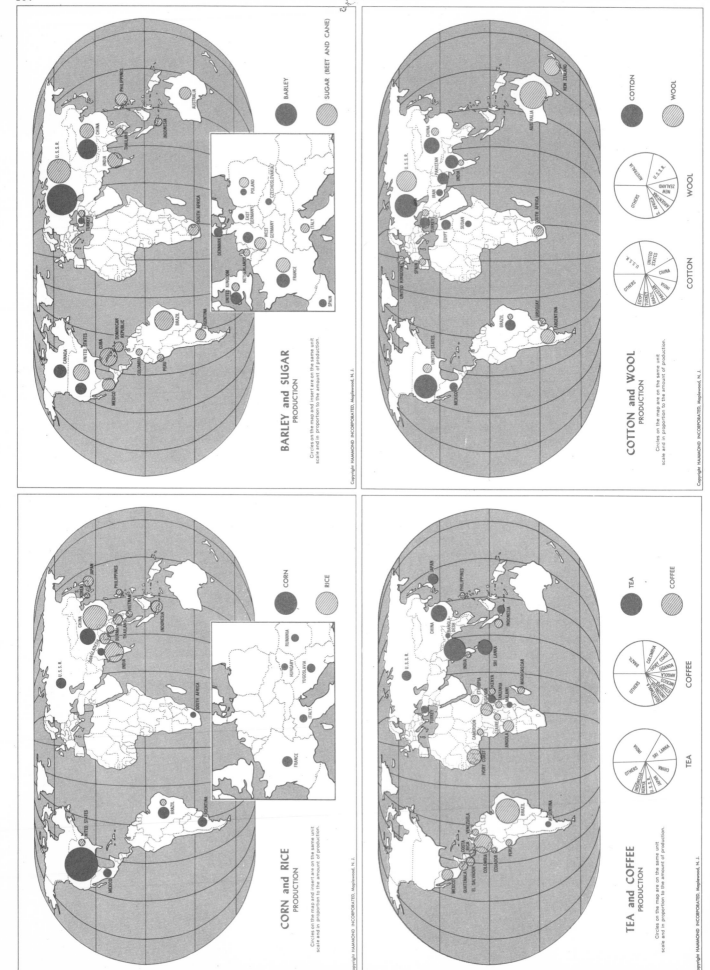

BARLEY and SUGAR
PRODUCTION

Circles on the map and insert are on the same unit scale and in proportion to the amount of production.

BARLEY   SUGAR (BEET AND CANE)

CORN and RICE
PRODUCTION

Circles on the map and insert are on the same unit scale and in proportion to the amount of production.

CORN   RICE

COTTON and WOOL
PRODUCTION

Circles on the map are on the same unit scale and in proportion to the amount of production.

COTTON   WOOL

WOOL

COTTON

TEA and COFFEE
PRODUCTION

Circles on the map are on the same unit scale and in proportion to the amount of production.

TEA   COFFEE

COFFEE

TEA

THE SOLAR SYSTEM

The *solar system* consists of the sun, nine planets and their 33 satellites, thousands of asteroids, millions of meteors and many comets. All these bodies travel around the sun in nearly circular paths called *orbits*. The planets are held in their orbits by the sun's gravity.

The nine planets surrounding the sun are commonly divided into two groups called the *inner planets* and the *outer planets*. The inner planets, those that are closest to the sun, include Mercury, Venus, Earth and Mars. Mercury is nearest the sun and, in this group, Mars is farthest from the sun. In the outer planets, or those farther from the sun, we find Jupiter, Saturn, Uranus, Neptune and Pluto.

The two planet groups are separated from one another by the *asteroid belt* which lies between Mars and Jupiter, forming the outer boundary of the inner group and the inner boundary of the outer planets.

Asteroids are small planets, most of which are only a few miles or less in diameter. The largest is Ceres, with a diameter of 480 miles, about 1/300 the size of the smallest planet, Mercury. Thousands of these small bodies go into making up the asteroid belt and it is worth noting that some scientists believe they once may have been one planet that was broken up.

In thinking of distances we tend to focus our attention on one dimension — the farther away something is the smaller it appears to be. Of course, with the solar system this is untrue. The largest of the planets, Jupiter, is in the outer planet group. The smallest planet, Mercury, is closest of all to the sun. To put these sizes in another perspective we should first look at the dimension of the sun.

In comparison with the Earth and the eight other planets in our solar system the sun is a gigantic sphere. In diameter it is approximately 864,000 miles, which is 100 times longer than Earth and 10 times longer than Jupiter, the biggest planet. In terms of volume, over 1½ million earth-size planets could fit into a sphere the size of the sun. But, although the sun is the largest body in our solar system, it is a relatively small *star* when viewed as part of our galaxy which contains more than 100 billion stars.

RELATIVE DIAMETERS OF THE PLANETS AND SUN

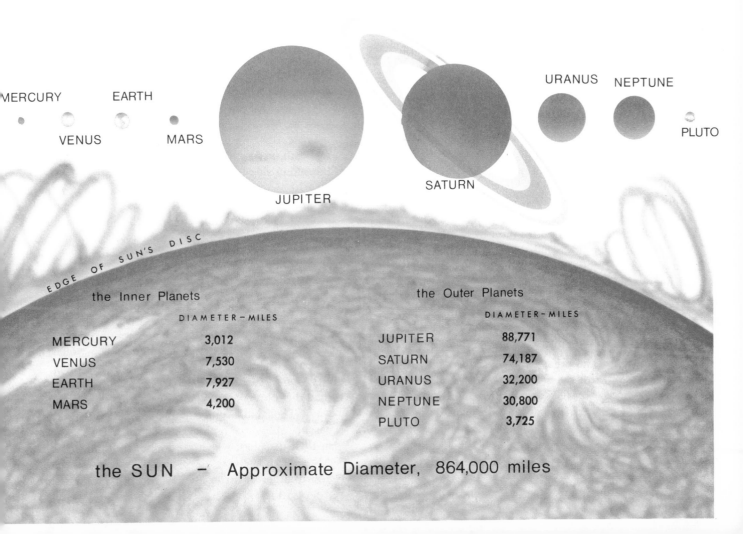

the Inner Planets	DIAMETER-MILES	the Outer Planets	DIAMETER-MILES
MERCURY	3,012	JUPITER	88,771
VENUS	7,530	SATURN	74,187
EARTH	7,927	URANUS	32,200
MARS	4,200	NEPTUNE	30,800
		PLUTO	3,725

the SUN – Approximate Diameter, 864,000 miles

THE SOLAR SYSTEM
(continued)

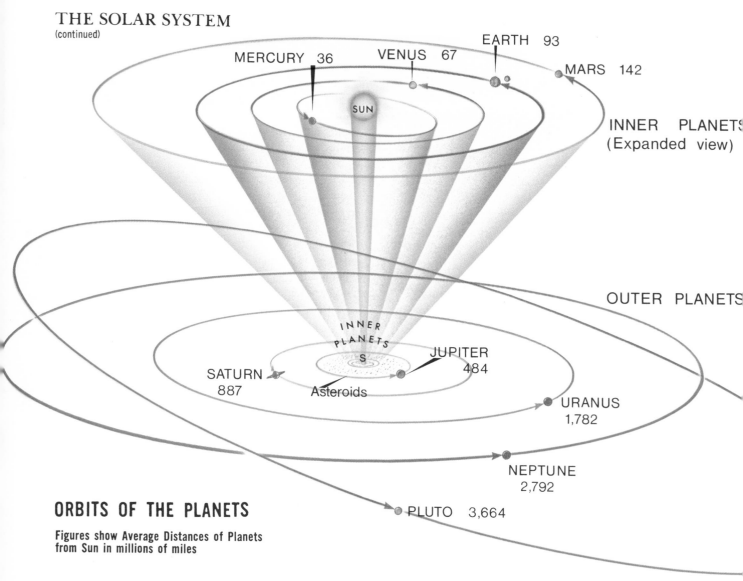

MERCURY 36   VENUS 67   EARTH 93   MARS 142

SUN

INNER PLANETS
(Expanded view)

OUTER PLANETS

INNER PLANETS

S

JUPITER
484

SATURN
887

Asteroids

URANUS
1,782

NEPTUNE
2,792

PLUTO 3,664

ORBITS OF THE PLANETS

Figures show Average Distances of Planets
from Sun in millions of miles

All the planets within our solar system move in the same direction around the sun and their orbits lie in nearly the same plane (within 3½ percent of the plane of the Earth's orbit). The orbits of Mercury and Pluto are exceptions to this. Mercury's orbit is inclined 7°; Pluto's orbit, inclined by over 17°, is highly eccentric. During part of its orbit Pluto lies within the orbit of Neptune. However the two planets do not collide because of the inclination of Pluto's orbit. Rotation and revolution are other factors of interest in studying planet characteristics. How much time does it take for each planet to rotate on its axis and to revolve around the sun? Mercury, for example, is closest to the sun and takes 88 days to make one revolution around the sun (the Earth takes 365¼ days or one earth-year). Pluto, on the other hand, is farthest from the sun and takes 248 earth-years to make one revolution. Pluto's axial rotation speed is also slow as compared with the 23 hours and 56 minutes it takes Earth to rotate once. It takes Pluto over six earth-days to make one rotation.

Temperature, weight (or gravity) and atmospheric density are other areas we study to understand our planet neighbors.

Mercury, for example, has no atmosphere. Its density

indicates that its composition is probably similar to that of Earth, yet its temperatures are a study in extremes. On the sunny side of Mercury temperatures may soar as high as, 770°F., while on the shadowed side the temperature approaches absolute zero (-459.6°F.). The gravitational pull on Mercury differs from Earth too. A man weighing 150 pounds on Earth would weigh 45 on Mercury.

Venus, closest to Earth, has a dense atmosphere composed mainly of carbon dioxide. Surface temperatures are considered to be quite hot.

Mars has an atmosphere considerably less than that of Earth so its surface is visible. Seasonal variations on Mars are indicated by the polar caps, believed to be ice and snow, which disappear in summer. This would indicate that Mars' atmosphere contains some water vapor.

Jupiter, Saturn, Uranus and Neptune are much larger than the other planets and differ from the others in several ways. Their bulk compositions include considerably more gaseous substance and less rock material. This is reflected in their densities, which are much less than that of Earth. Also, they have extremely low surface temperatures. Relatively little is known about *Pluto* because it is so distant from Earth.

A MAN'S WEIGHT AND JUMP ON EACH PLANET

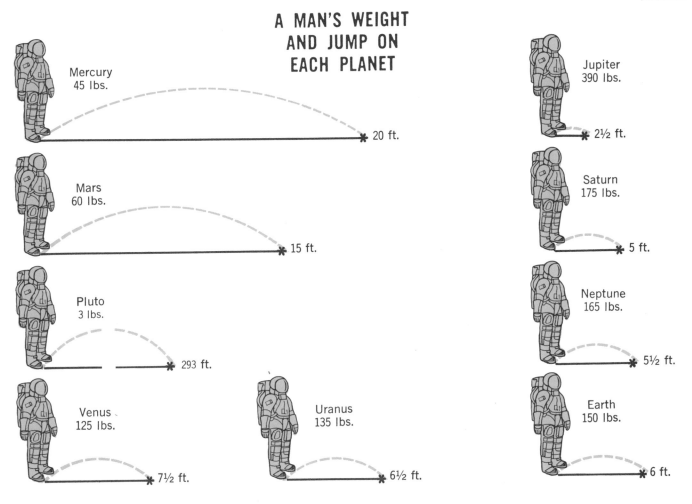

Mercury
45 lbs.
20 ft.

Mars
60 lbs.
15 ft.

Pluto
3 lbs.
293 ft.

Venus
125 lbs.
7½ ft.

Uranus
135 lbs.
6½ ft.

Jupiter
390 lbs.
2½ ft.

Saturn
175 lbs.
5 ft.

Neptune
165 lbs.
5½ ft.

Earth
150 lbs.
6 ft.

FACTS ABOUT THE PLANETS

	MERCURY	VENUS	EARTH	MARS	JUPITER	SATURN	URANUS	NEPTUNE	PLUTO
Period of Revolution Around the Sun	87.97 days	224.7 days	365.26 da.	687 days	11.86 years	29.46 years	84.02 years	164.79 yrs.	248.5 years
Period of Rotation on Axis	59 days	247 days	23 hours 56 min.	24 hours 37 min.	9 hours 50 min.	10 hours 14 min.	10 hours 45 min.	15 hours 48 min.	6 days 9 hours ?
Inclination of Axis	?	?	23.5°	24°	3.1°	26.8°	98°	29°	?
Minimum Distance from Earth (millions of miles)	49	26	—	34	362	773	1,594	2,654	2,605
Maximum Distance from Earth (millions of miles)	137	161	—	247	597	1,023	1,946	2,891	4,506
Escape Velocity (miles per hour)	7,920	22,700	25,200	11,200	133,200	79,200	46,800	55,400	?
Mass or Weight (compared to Earth)	.05	.81	1.0	.11	318	95	14.6	17.2	.0013?
Volume (compared to Earth)	.06	.92	1.0	.15	1,318	736	64	60	.016?
Density (water = 1.0)	5.5	5.25	5.5	4.0	1.3	0.7	1.3	1.6	4.5
Number of Moons	0	0	1	2	15	10	5	2	1

THE MOON: EARTH'S NATURAL SATELLITE

The moon, man's first stepping stone into the silent seas of space, actually is a gigantic stone in the sky ... an airless, waterless sphere of towering mountain ranges, broad craters, great plains and powdery, gray-brown dust.

It rotates around the earth keeping its far side always hidden from our sight. One-quarter the diameter of the earth, and having one-sixth its gravity, this uninviting neighbor only 238,000 miles away was formed, we speculate, when

NEAR SIDE

(photo mosaic)

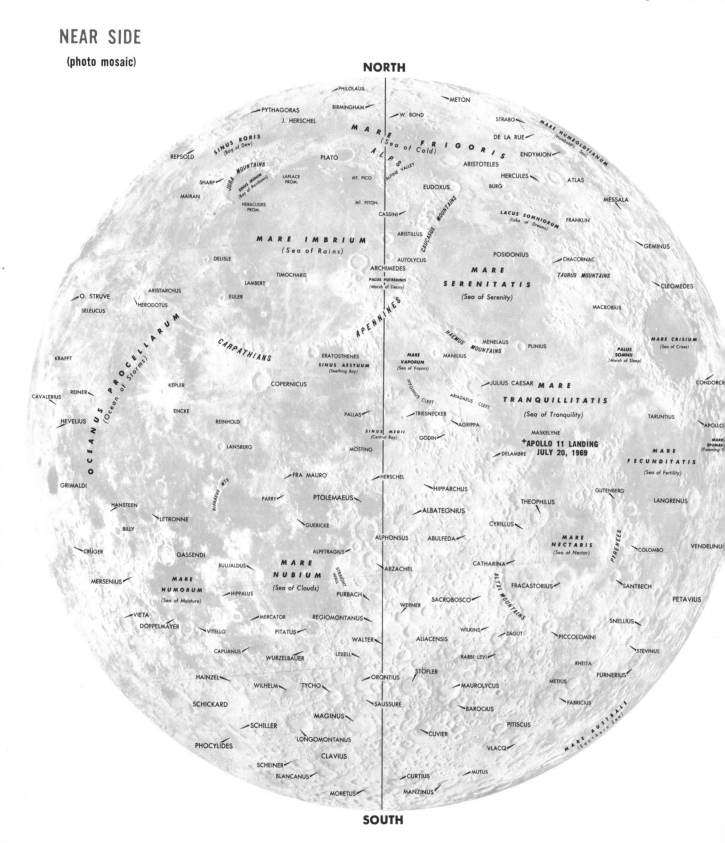

a swirling cloud of cosmic gas and particles separated into eddies which contracted to become the sun, the planets and their satellites. Unshielded by protective air, temperatures on its surface range from over 200°F. by day to −200°F. after dark. Yet some day we shall launch space vehicles from there, virtually without gravitational drag, and, because it has no atmosphere, clearly observe the furthest heavens on ever-cloudless nights.

FAR SIDE

(artistic rendering of Lunar Orbiter photographs — NASA)

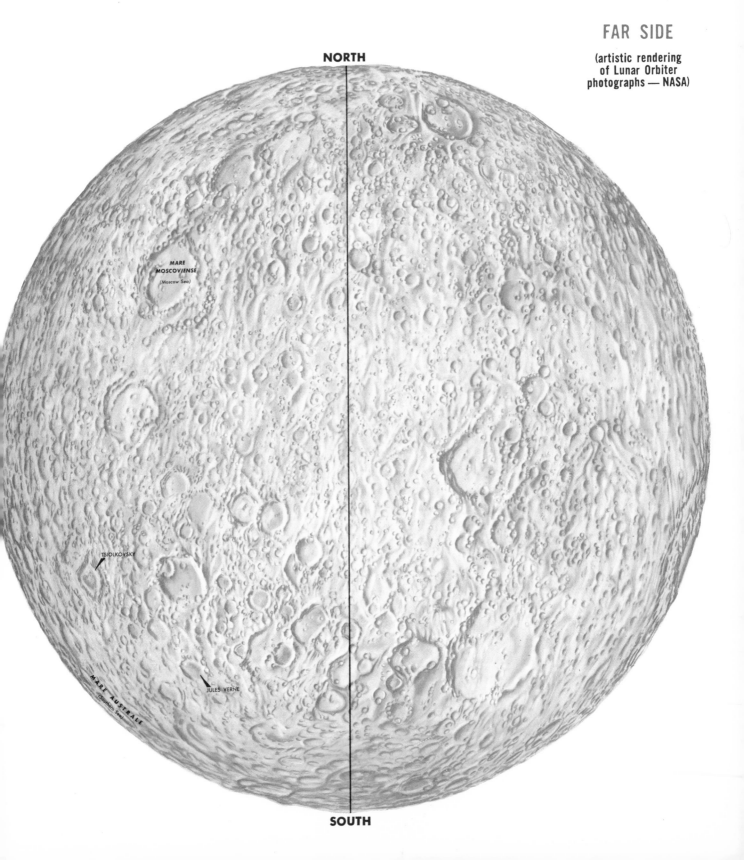

NORTH

MARE MOSCOVIENSE (Moscow Sea)

TSIOLKOVSKY

JULES VERNE

MARE AUSTRALE (Southern Sea)

SOUTH

DEVELOPMENT OF CONTINENTS AND OCEANS

If we can envision the continents of the world as seated firmly on massive rafts of rock and moving across the surface of the earth at a rate of about 6 feet every 60 years we have a basic notion of what is meant by continental drift and the manner in which land and sea masses have been formed.

The original concept of continental drift was proposed in the 1920s, but only during the past three years or so have geologists and geophysicists accepted as fact the seemingly preposterous notion that the surface of the earth is constantly in motion.

The making of the continents began more than 200 million years ago during the Permian period with the splitting of a gigantic landmass known as Pangaea. Two continents, Laurasia to the north and Gondwana to the south, were formed by the initial division. Over a period of many millions of years these landmasses subdivided into smaller parts approximately the shapes of Africa, Eurasia, North and South America, Australia, and Antarctica as we know them today.

CONTINENTAL DRIFT

Source: R. S. Dietz and J. C. Holden

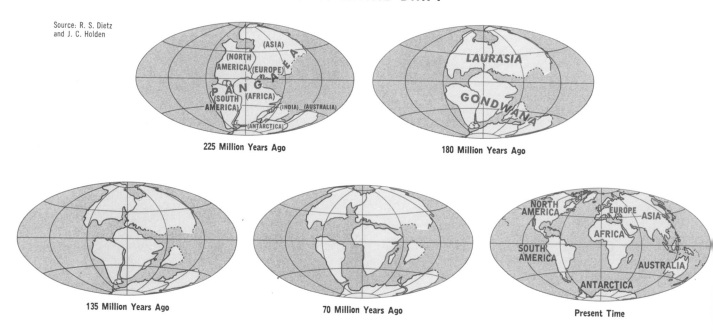

225 Million Years Ago

180 Million Years Ago

135 Million Years Ago

70 Million Years Ago

Present Time

CRUSTAL MOVEMENT

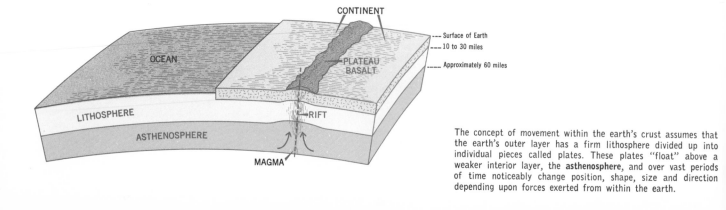

The concept of movement within the earth's crust assumes that the earth's outer layer has a firm lithosphere divided up into individual pieces called plates. These plates "float" above a weaker interior layer, the **asthenosphere**, and over vast periods of time noticeably change position, shape, size and direction depending upon forces exerted from within the earth.

THE ICE AGES

Far from being an isolated instance, the movement of glaciers over the face of the earth has been a natural phenomenon for many thousands of years. Stimulated by changes in climate and resulting changes in sea level — perhaps induced by shifts in the earth's axis — glaciers have followed a rather unpredictable course of advance and retreat continuing into the 20th century.

At some point in unrecorded history during the greatest ice age, or the Pleistocene epoch, as much as 27 percent of the earth's surface was covered by glacial ice to a depth of up to 10,000 feet. The icy masses moved across the earth as far south as New York City and the Missouri River in North America, burying much of Europe and blanketing vast areas in northern Asia.

Many of the great ice sheets retreated as the climate became warmer, leaving deposits of soil and rock picked up as they traveled southward in the Northern Hemisphere. The landscape changed as the glaciers left behind their typical U-shaped valleys, amphitheater-like hollows and jagged mountain ridges, altering to a large extent the former ecological zones which changed again and again as the ice reformed and melted.

Although not enough is known about glaciers to predict accurately their future behavior, we do know that they react to climatic changes. Glaciers were advancing in Alpine regions during the 19th century until a global warm up in the beginning of this century caused their retreat. Recently the trend has been toward cooler and moister climate and, on a limited scale, glaciers are beginning to advance once more.

EXTENT OF GLACIATION
IN THE NORTHERN HEMISPHERE
DURING THE ICE AGES

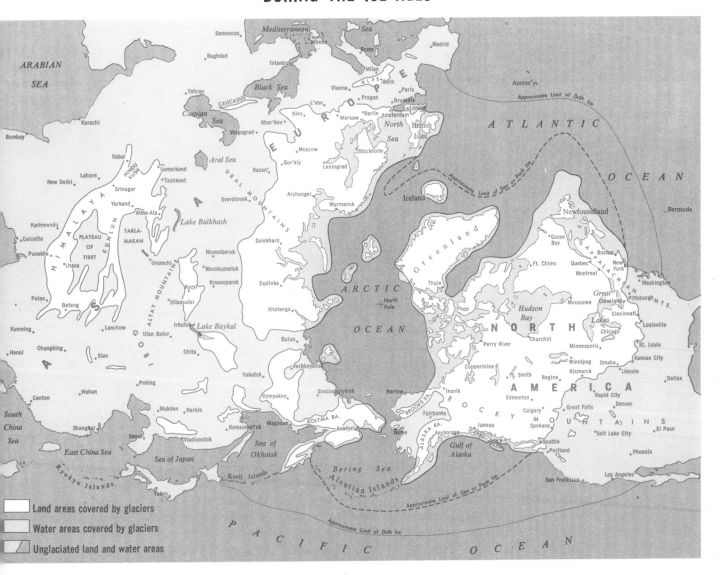

Land areas covered by glaciers

Water areas covered by glaciers

Unglaciated land and water areas

THE GEOLOGIC RECORD

GEOLOGIC TIME

TIME DIVISION			YEARS AGO	MAJOR GEOLOGIC DEVELOPMENTS
CENOZOIC ERA	QUATERNARY PERIOD	RECENT	10,000	GREAT LAKES NORWEGIAN FJORDS ICE AGES BLACK SEA
		PLEISTOCENE	1-2 million	
	TERTIARY PERIOD	PLIOCENE	11 million	CASPIAN SEA
		MIOCENE	25 million	HIMALAYAS
		OLIGOCENE	40 million	ALPS
		EOCENE	60 million	
		PALEOCENE	70 million	ANDES MOUNTAINS ROCKY MOUNTAINS CHALK DEPOSITS
MESOZOIC ERA		CRETACEOUS PERIOD	135 million	COAST RANGES SIERRA NEVADA JURA MOUNTAINS
		JURASSIC PERIOD	180 million	NEW JERSEY PALISADES
		TRIASSIC PERIOD	225 million	CAUCASUS URAL MOUNTAINS APPALACHIAN MOUNTAINS
PALEOZOIC ERA		PERMIAN PERIOD	270 million	POTASH DEPOSITS
		PENNSYLVANIAN PERIOD	300 million	COAL DEPOSITS
		MISSISSIPPIAN PERIOD	350 million	ACADIAN MOUNTAINS
		DEVONIAN PERIOD	400 million	
		SILURIAN PERIOD	440 million	NIAGARA FALLS CAPROCK TACONIC MOUNTAINS
		ORDOVICIAN PERIOD	500 million	LIMESTONE DEPOSITS VERMONT MOUNTAINS
		CAMBRIAN PERIOD	600 million	ARIZONA MOUNTAINS
		PRE-CAMBRIAN		METALLIC ORE DEPOSITS LAURENTIAN MOUNTAINS ADIRONDACK MOUNTAINS

Like a giant Rosetta stone the secrets of the earth's creation lie spread in strata beneath our feet, revealing their hieroglyphic message to a few of the initiated.

For billions of years layers of rock — the sedimentary deposits of ages — have piled up on the earth's surface, entrapping the characteristics of time. Time when a lifeless nature prepared for the first microscopic living organisms; time when these organisms were destroyed or became extinct, time when, through endless subtle mutations, they evolved into new forms of life.

The Paleozoic, ancient era; Mesozoic, middle era; and

Cenozoic, recent era, are the designations used for the broad periods of time during which life evolved. Locked within strata of rock, vestiges of life are found in the fossilized remains of creatures over a billion years old. In succeeding layers geologists and anthropologists find other clues to the mystery of time and life: the appearance of the lowest forms of animal life; the evolution of fish, amphibians, reptiles, birds and mammals. Late in the schedule of creation traces of a strange and wonderful animal appear, for it was only one million years ago that man left his first imprint on the geologic record.

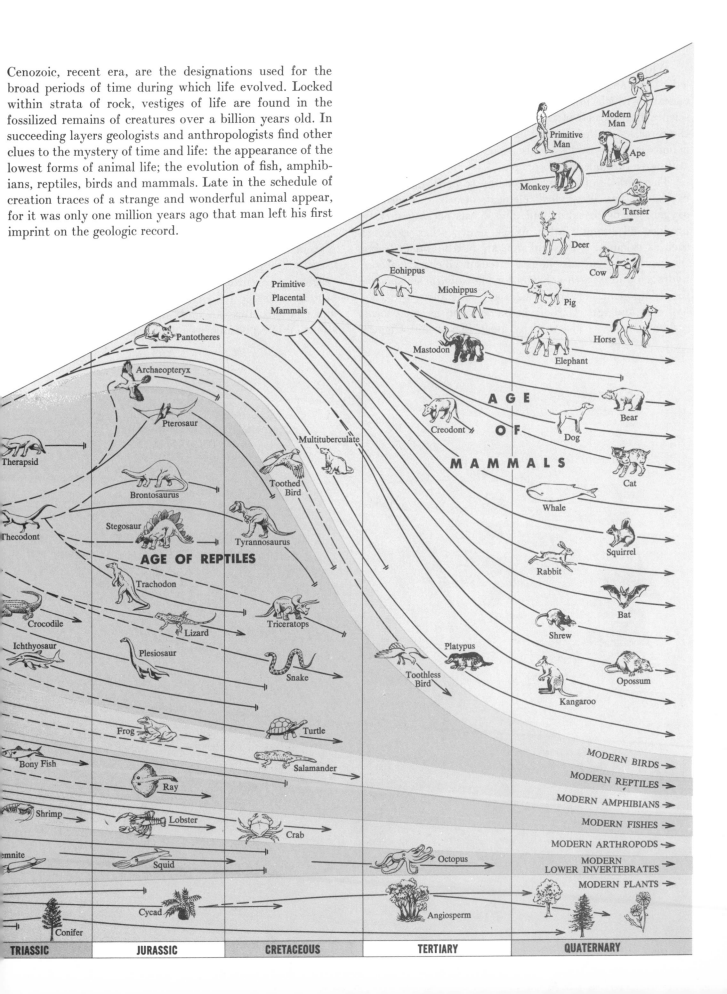

LIFE SUPPORT CYCLES

With an intuition clearly beyond their scientific knowledge, the ancients of India developed a theory of reincarnation which, in some philosophic ways, parallels what science has learned of the workings of the biosphere. In the remarkable thrift of nature nothing is lost — in tremendous complex cycles atoms from the first life on earth still move through the biosphere.

The miracle of energy is constantly performed in the cycles of the "life-giving" elements. Carbon, hydrogen, oxygen, nitrogen, sulfur and phosphorus act together to produce all living matter. While many other elements such as calcium, iodine and iron are also found in living things, they are not absolute essentials in all cases. Carbon, hydrogen and oxygen are vital for photosynthesis and are the components of the basic food substances — carbohydrates and fats. Carbon, in its common gaseous form, carbon dioxide, is absorbed by green plants and triggers

the production of carbohydrate compounds by reacting with molecules of water.

Some "energy" is stored within the plant in the form of new tissue; other "energy," in the form of oxygen is released into the air to be used by other organisms. The seemingly inexhaustible supply of carbon dioxide available for use is replenished in the atmosphere through the respiration of all living things, and in the soil as bacteria and fungi break down plant and animal cells,

Nitrogen, sulfur and phosphorus are essential to animals and plants for the production and maintenance of protein. Nitrogen, with carbon, hydrogen and oxygen, is used for the growth and repair of tissue. Sulfur acts as a "stiffening" agent in all protein. To perform their functions proteins must be folded and shaped in a particular way, and their structure is maintained by bonds between sulfur atoms. While phosphorus is not a constituent of protein,

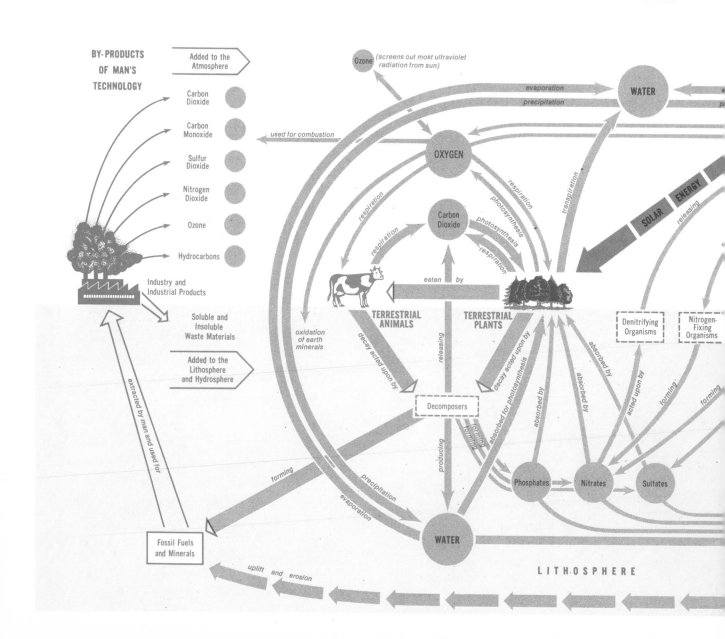

no protein can be made without it. Special phosphate compounds are the "fuel" for all biochemical work within the cell.

Although about four-fifths of the atmosphere is nitrogen, higher forms of life cannot make use of it in its "free" state and must absorb it at one or more points in its biospheric cycle. The decomposers — bacteria and fungi — act on waste matter, breaking down complex compounds into simpler usable forms including nitrogen. Some nitrogen-fixing bacteria are able to utilize atmospheric nitrogen in their own metabolism, while others convert it to those nitrogen-enriched substances necessary for all plant growth.

In nature, no part is greater than the whole and almost every element is dependent on another for some essential part of its cycle. Water, which is incorporated into every organism, essential in the formation of free oxygen which in turn sustains the life of that organism. Water is also

the principal "carrier" in the cycling of all elements. When it evaporates, water returns certain elements to the atmosphere; when it seeps through the soil on its return to the sea, water distributes nutrients to plant roots.

Carbon monoxide, sulfur and nitrogen oxides, hydrocarbons — by-products of man's industry — are being injected into the biosphere in ever-increasing amounts. There, as the "new compounds," they must in some way co-exist with the life-support cycles established throughout millions of years of evolution. Their compatability with these cycles and the organisms they nurture will determine the future of life on our planet.

Already man has learned one thing. Although the question of reincarnation or any form of life after death remains unanswered for many, science has proved that there is no natural end to the raw materials of nature or to the "new compounds" man has made from them.

INTERLOCKING CYCLES OF THE BIOSPHERE

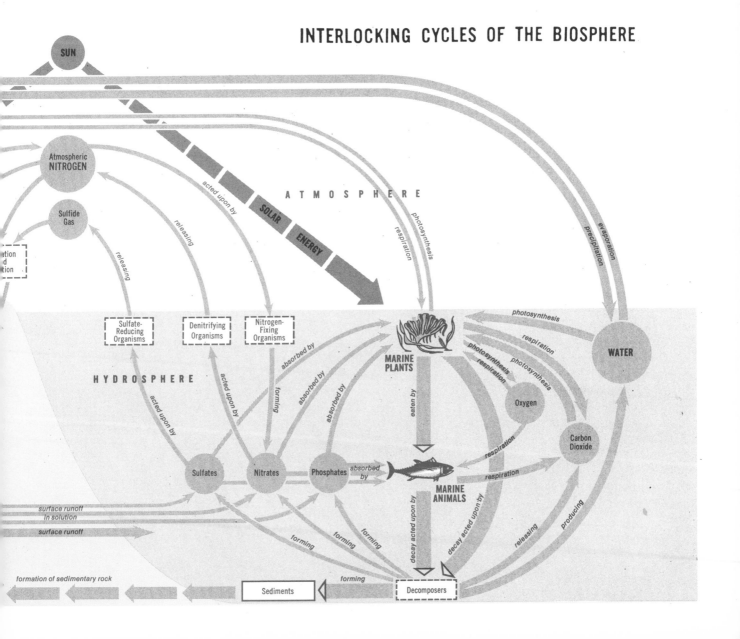

MAN'S IMPACT UPON NATURE

Since he could think man has been at war with death. He has fought his battles against destruction with science and technology as his weapons, virtually eliminating his own annihilation by predatory animals and from diseases such as leprosy, tuberculosis and diphtheria. He has walked into many valleys of death to fight malaria and yellow fever, and he has resolved that each year more of his own kind will live to finish out their threescore years and ten.

However, the victory over nature, which had balanced population with food supply and space, is bitter, for the population has "exploded" leaving man with the seemingly insolvable problem of providing more food and space for himself or reducing his numbers by starvation or by war.

Man outsmarted himself in many ways as he worked toward creating a more perfect world for himself without understanding that natural laws go beyond human manipulation. He has destroyed forests and meadows, polluted the water and air, eliminated organisms that tried to share his bread. However, he has yet to learn to recreate the wood and brush or the interdependent communities of

bacteria, insects and animals that he learned — too late — enrich the air, the soil and the water and without which he cannot function.

Modern man knows how to manufacture "miraculous" materials to work for his pleasure or his seemingly insatiable needs, but the sophistications of technology have yet to control effectively the by-products. These new materials, still subject to the order of nature's cycles, penetrate the biosphere and eventually come to roost in his own vulnerable body.

New battles are being fought throughout the world and new standards bearing the slogans of ecology float in the "unsafe" air. It is somehow ironic to find that many people now believe that man has been fighting the wrong fight in his gigantic struggle with nature. That, after all, nature never was his enemy.

Man cannot turn back to his beginnings when he lived with, and not against, the natural world. But a compromise between technology and nature must take place for our "plundered planet" cries out for the day of reckoning.

POLLUTION CIRCLE

TYPES OF POLLUTION AND THEIR EFFECT ON THE TOTAL ENVIRONMENT

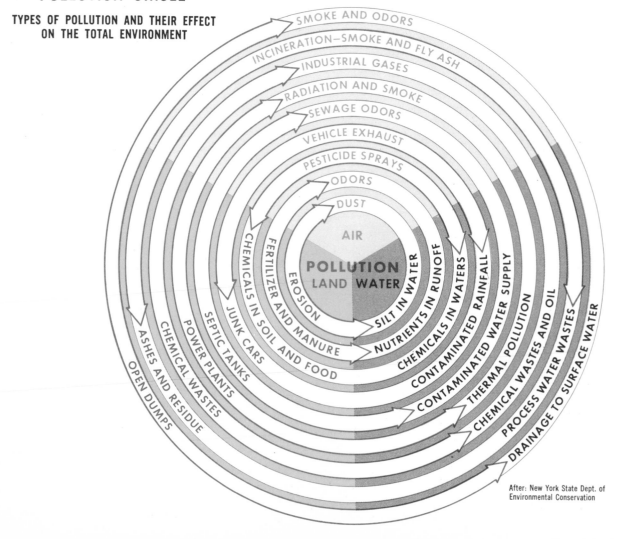

After: New York State Dept. of Environmental Conservation

WORLD HISTORY SECTION

CONTENTS

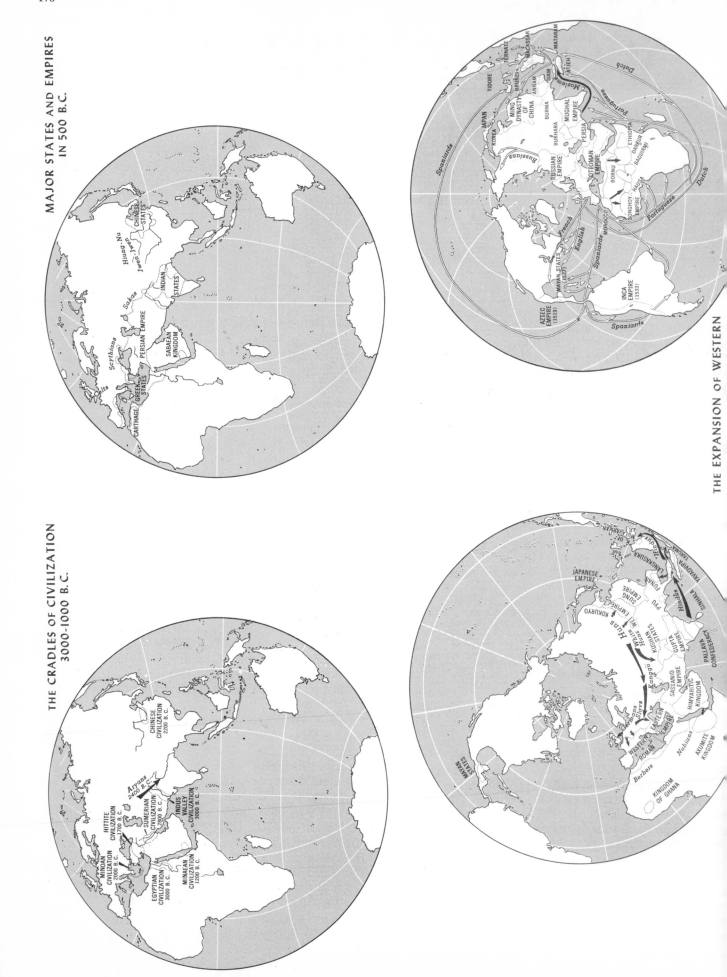

MAJOR STATES AND EMPIRES
IN 500 B.C.

THE CRADLES OF CIVILIZATION
3000-1000 B.C.

THE EXPANSION OF WESTERN

ANCIENT SEMITIC WORLD

Copyright by C. S. HAMMOND & Co., N. Y.

Scale of Miles
0 — 100 — 200 — 300 — 400

Black Sea
Caspian Sea
ASIA MINOR
Hittite Kingdom
Lycia
Cilicia
Kittim (Cyprus)
Mediterranean Sea
Mt. Ararat
Assyria
Media
Mitanni
Haran
Hatti
Carchemish
Nineveh
Asshur
Arvad
Hamath
ARAM (Syria)
Amurru
Naharina
Euphrates R.
Tigris R.
Akkad
Elam
Sidon
Damascus
Tyre
Babylon
BABYLONIA
Sumer
Susa
Jerusalem
Ammon
Moab
Ur
Chaldea
Persian Gulf
EGYPT
Memphis
Mizraim
Nile R.
Pathros (Upper Egypt)
Mt. Sinai
ARABIA
Arabian Desert
Red Sea
Thebes
Cush
Ethiopia

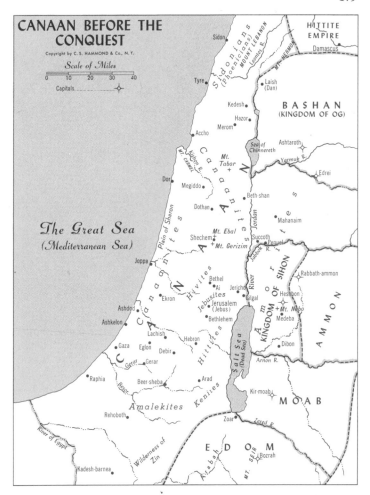

CANAAN BEFORE THE CONQUEST

Copyright by C. S. HAMMOND & Co., N. Y.

Scale of Miles
0 — 20 — 40

Capitals

HITTITE EMPIRE
Sidon
Sidonians (Phoenicians)
MOUNT LEBANON
Leontes R.
MT. HERMON
Damascus
Tyre
Laish (Dan)
Kedesh
BASHAN (KINGDOM OF OG)
Hazor
Accho
Merom
Ashtaroth
Sea of Chinnereth
Yarmuk R.
Edrei
Dor
Mt. Tabor
Megiddo
Beth-shan
Canaanites
Dothan
Mahanaim
Succoth
Peniel
Jordan River
The Great Sea (Mediterranean Sea)
Joppa
Shechem
Mt. Ebal
Mt. Gerizim
Hivites
Bethel
Ai
Jericho
Gilgal
Rabbath-ammon
Heshbon
KINGDOM OF SIHON
Mt. Nebo
AMMON
Ekron
Jebusites
Jerusalem (Jebus)
Ashdod
Bethlehem
Medeba
Ashkelon
Hittites
Salt Sea (Dead Sea)
Gaza
Lachish
Hebron
Eglon
Debir
Dibon
Gerar
Gerar
Arad
Kir-moab
Raphia
Beer-sheba
Amalekites
Kenites
MOAB
Rehoboth
Zoar
Zered R.
River of Egypt
Wilderness of Zin
EDOM
Bozrah
MT. SEIR
Kadesh-barnea

CANAAN AS DIVIDED AMONG THE TWELVE TRIBES
c. 1200-1020 B.C.

Copyright by C. S. HAMMOND & Co., N. Y.

Scale of Miles
0 — 10 — 20 — 30 — 40

Sidon
Sidonians (Phoenicians)
MOUNT LEBANON
Leontes R.
MT. HERMON
Zarephath
Damascus
Tyre
Kanah
DAN
Laish (Dan)
ASHER
NAPHTALI
Bashan
Kedesh
Accho
Cabul
Hazor
MANASSEH
Mt. Carmel
Sea of Chinnereth
Ashtaroth
Dor
ZEBULUN
Hammath
Mt. Tabor
Aphek
Megiddo
ISSACHAR
Shunem
Yarmuk R.
Edrei
Taanach
Jezreel
Ramoth-gilead
Beth-shan
Plain of Sharon
MANASSEH
Mahanaim
GAD
AMMON
Mt. Ebal
Jordan River
Mt. Gerizim
Shechem
Succoth
Jabbok R.
Peniel
The Great Sea (Mediterranean Sea)
EPHRAIM
Shiloh
Jazer
Joppa (Japho)
Aijalon
Beth-horon
Bethel
Rabbath-ammon
Jabneel
Gibeon
Jericho
DAN
Gezer
BENJAMIN
Gilgal
Ashdod
Jerusalem (Jebus)
Heshbon
Mt. Nebo
Libnah
Bethshemesh
Bethlehem
Medeba
Ashkelon
JUDAH
Philistines
Salt Sea (Dead Sea)
REUBEN
Gaza
Lachish
Hebron
Caleb
Gerar
Ziklag
En-gedi
Aroer
Raphia
Cherethites
Beer-sheba
Kenites
Arnon R.
Hormah
SIMEON
MOAB
Rehoboth
Zered R.
River of Egypt
Wilderness of Zin
EDOM

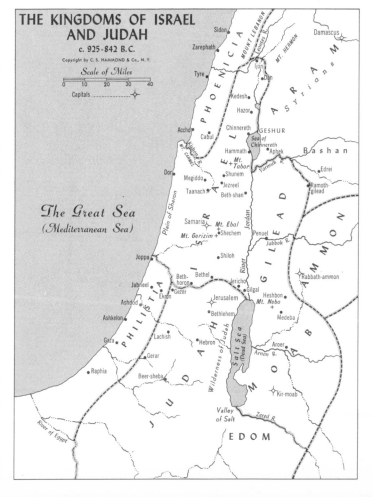

THE KINGDOMS OF ISRAEL AND JUDAH
c. 925-842 B.C.

Copyright by C. S. HAMMOND & Co., N. Y.

Scale of Miles
0 — 10 — 20 — 30 — 40

Capitals

Sidon
PHOENICIA
MOUNT LEBANON
Leontes R.
MT. HERMON
Damascus
Zarephath
Tyre
Ijon
Dan
ARAM
Kedesh
Syrians
Hazor
Accho
Chinnereth
GESHUR
Cabul
Sea of Chinnereth
Aphek
Bashan
Dor
Hammath
Mt. Tabor
Yarmuk R.
Megiddo
Shunem
Jezreel
Edrei
Taanach
Beth-shan
Ramoth-gilead
Plain of Sharon
ISRAEL
Mahanaim
GILEAD
AMMON
Samaria
Mt. Ebal
Jordan River
Mt. Gerizim
Shechem
Penuel
Jabbok R.
The Great Sea (Mediterranean Sea)
Shiloh
Joppa
Bethel
Rabbath-ammon
Beth-horon
Jabneel
Jericho
Ekron
Gezer
Gilgal
Ashdod
Jerusalem
Heshbon
Mt. Nebo
Bethlehem
Medeba
Ashkelon
PHILISTIA
JUDAH
Wilderness of Judah
Salt Sea (Dead Sea)
MOAB
Gaza
Lachish
Hebron
Gerar
Aroer
Raphia
Beer-sheba
Arnon R.
Kir-moab
Valley of Salt
River of Egypt
EDOM

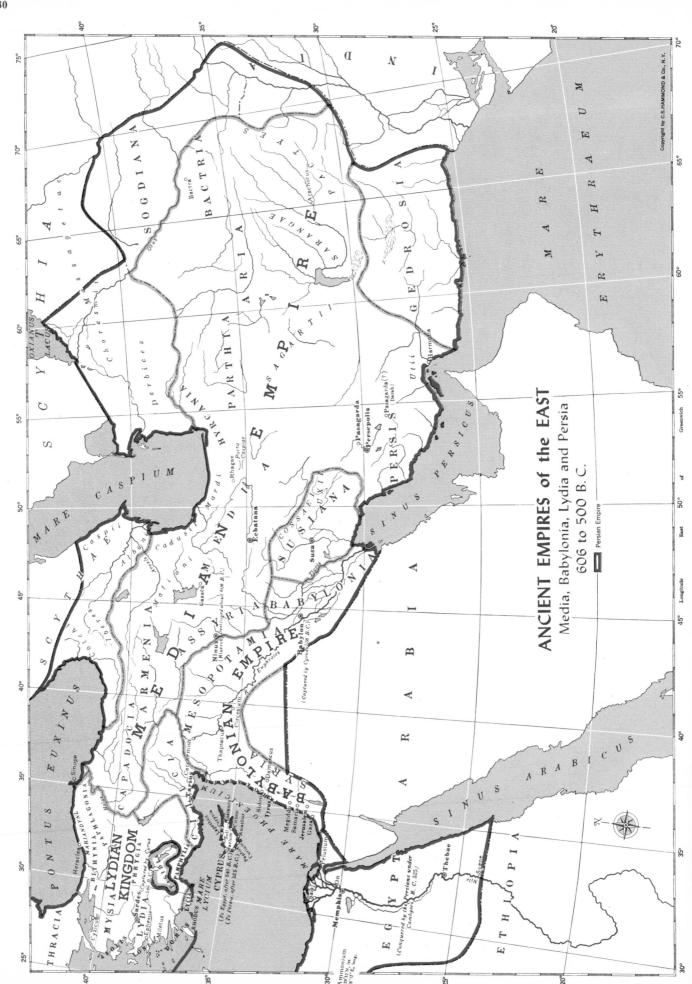

ANCIENT EMPIRES of the EAST
Media, Babylonia, Lydia and Persia
606 to 500 B.C.

Persian Empire

Copyright by C.S.HAMMOND & Co., N.Y.

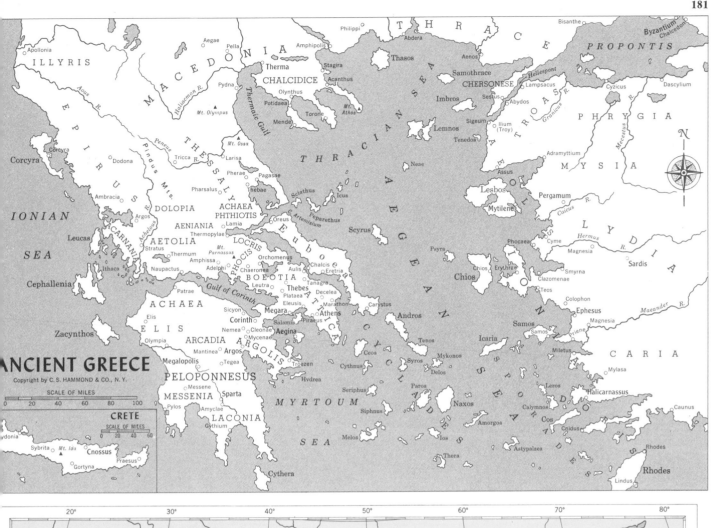

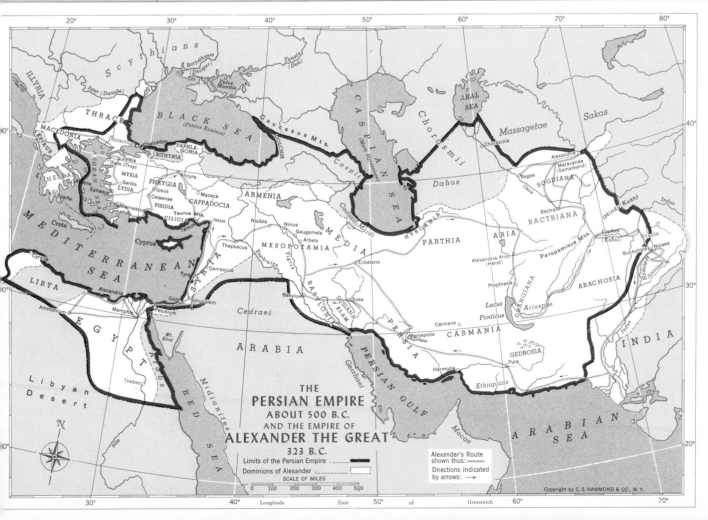

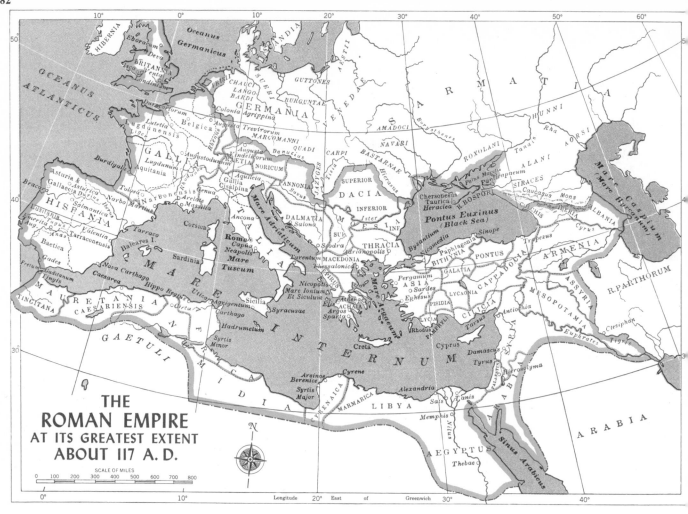

THE ROMAN EMPIRE
AT ITS GREATEST EXTENT
ABOUT 117 A. D.

SCALE OF MILES
0 100 200 300 400 500 600 700 800

Longitude 20° East of Greenwich 30°

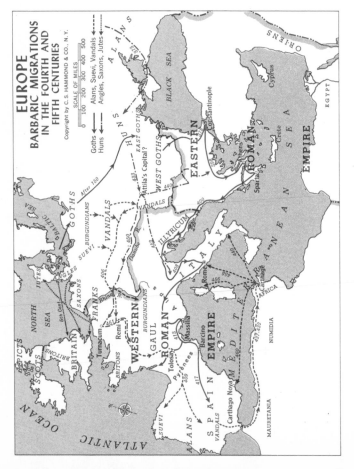

EUROPE
BARBARIC MIGRATIONS
IN THE FOURTH AND
FIFTH CENTURIES

Copyright by C. S. HAMMOND & CO., N.Y.

SCALE OF MILES
0 100 200 300 400 500

⇢ Alans, Suevi, Vandals
⇠ Angles, Saxons, Jutes
→ Goths
→ Huns

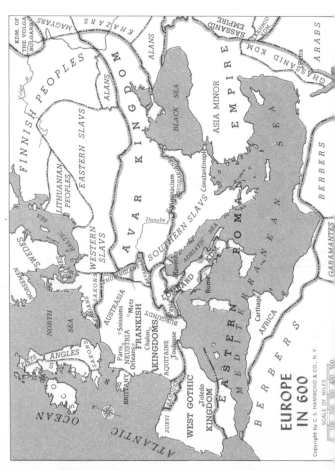

EUROPE
IN 600

Copyright by C. S. HAMMOND & CO., N.Y.

SCALE OF MILES

BRITANNIA
about 350 A.D
Showing the
CELTIC TRIBES
and approximately
The 4 Divisions of DIOCLETIAN
SCALE OF MILES
0 20 40 60 80 100

ENGLISH CONQUEST
From 450 to the End of the 6th Century
Showing the Settlements of the Jutes,
Saxons and Angles. Also the Sections
of the Country which were retained by
the Britons (Celtic Tribes).
SCALE OF MILES
0 20 40 60 80 100

ENGLAND
in the Eighth Century
(The "HEPTARCHY")
SCALE OF MILES
0 20 40 60 80 100

ENGLAND
after the Peace of Wedmore
(878 A.D.)
Showing the Divisions between
ALFRED and GUTHRUM
SCALE OF MILES
0 20 40 60 80 100

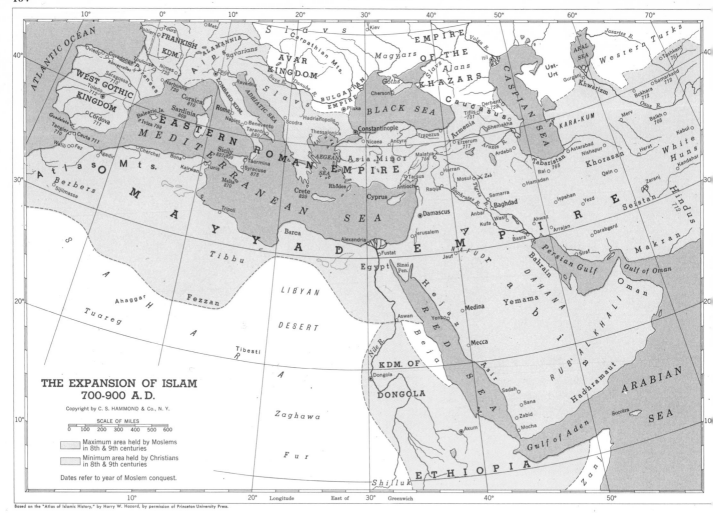

THE EXPANSION OF ISLAM
700-900 A.D.

Copyright by C. S. HAMMOND & Co., N. Y.

SCALE OF MILES
0 100 200 300 400 500 600

Maximum area held by Moslems
in 8th & 9th centuries

Minimum area held by Christians
in 8th & 9th centuries

Dates refer to year of Moslem conquest.

Based on the "Atlas of Islamic History," by Harry W. Hazard, by permission of Princeton University Press.

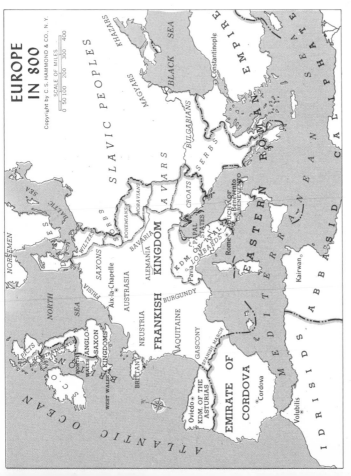

EUROPE
IN 800

Copyright by C. S. HAMMOND & Co., N. Y.

SCALE OF MILES
0 50 100 200 300 400

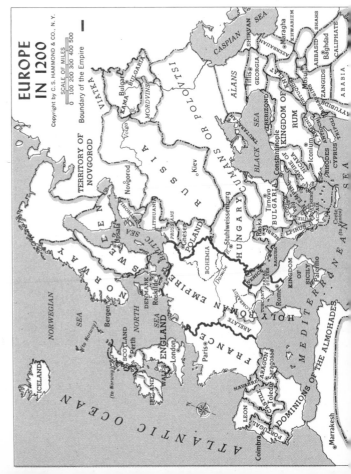

EUROPE
IN 1200

Copyright by C. S. HAMMOND & Co., N. Y.

SCALE OF MILES
0 100 200 300 400 500

Boundary of the Empire

ENGLISH POSSESSIONS IN FRANCE

Possessions of William the Conqueror:
Possessions of Henry II, about 1180:
Possessions of Henry III, 1272:
French Crown Lands, 1180:
Boundary of France in the 12th Century:

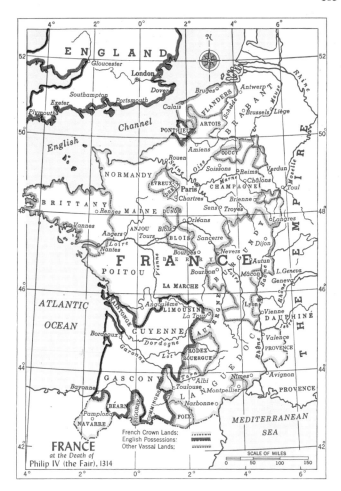

FRANCE
at the Death of
Philip IV (the Fair), 1314

French Crown Lands:
English Possessions:
Other Vassal Lands:

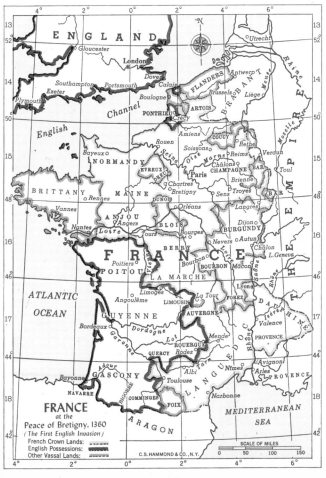

FRANCE
at the
Peace of Bretigny, 1360
(The First English Invasion)

French Crown Lands:
English Possessions:
Other Vassal Lands:

C.S. HAMMOND & CO., N.Y.

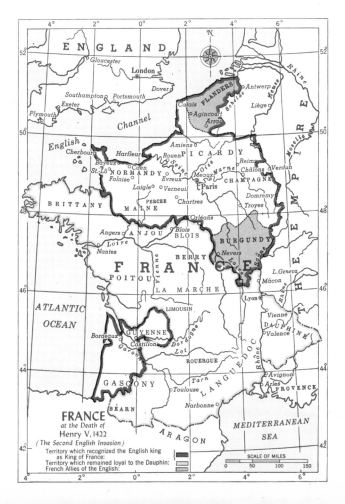

FRANCE
at the Death of
Henry V, 1422
(The Second English Invasion)

Territory which recognized the English king
as King of France:
Territory which remained loyal to the Dauphin:
French Allies of the English:

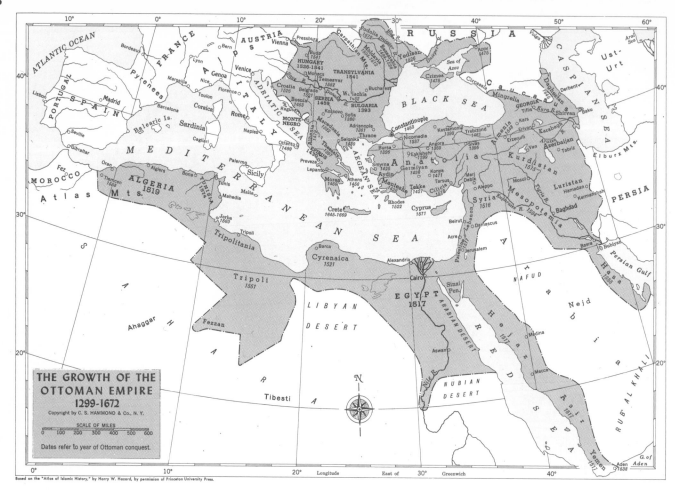

**THE GROWTH OF THE
OTTOMAN EMPIRE
1299-1672**

Copyright by C. S. Hammond & Co., N. Y.

SCALE OF MILES

0 100 200 300 400 500 600

Dates refer to year of Ottoman conquest.

Based on the "Atlas of Islamic History," by Harry W. Hazard, by permission of Princeton University Press.

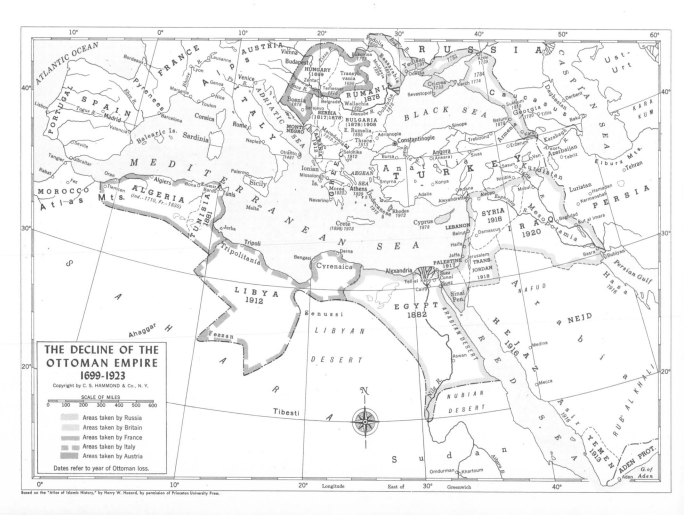

**THE DECLINE OF THE
OTTOMAN EMPIRE
1699-1923**

Copyright by C. S. Hammond & Co., N. Y.

SCALE OF MILES

0 100 200 300 400 500 600

Areas taken by Russia
Areas taken by Britain
Areas taken by France
Areas taken by Italy
Areas taken by Austria

Dates refer to year of Ottoman loss.

Based on the "Atlas of Islamic History," by Harry W. Hazard, by permission of Princeton University Press.

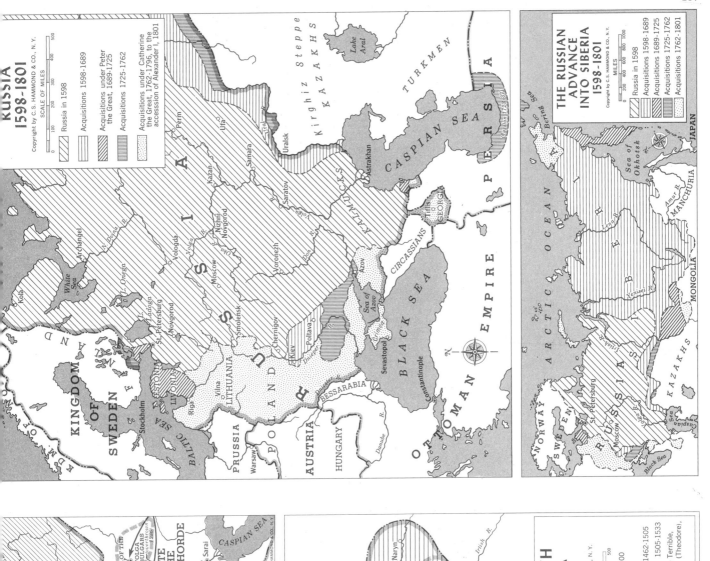

RUSSIA 1598-1801

Copyright by C.S. HAMMOND & CO., N.Y.

SCALE OF MILES

	Russia in 1598
	Acquisitions 1598-1689
	Acquisitions under Peter the Great, 1689-1725
	Acquisitions 1725-1762
	Acquisitions under Catherine the Great, 1762-1796, to the accession of Alexander I, 1801

THE RUSSIAN ADVANCE INTO SIBERIA 1598-1801

Copyright by C.S. HAMMOND & CO., N.Y.

MILES

	Russia in 1598
	Acquisitions 1598-1689
	Acquisitions 1689-1725
	Acquisitions 1725-1762
	Acquisitions 1762-1801

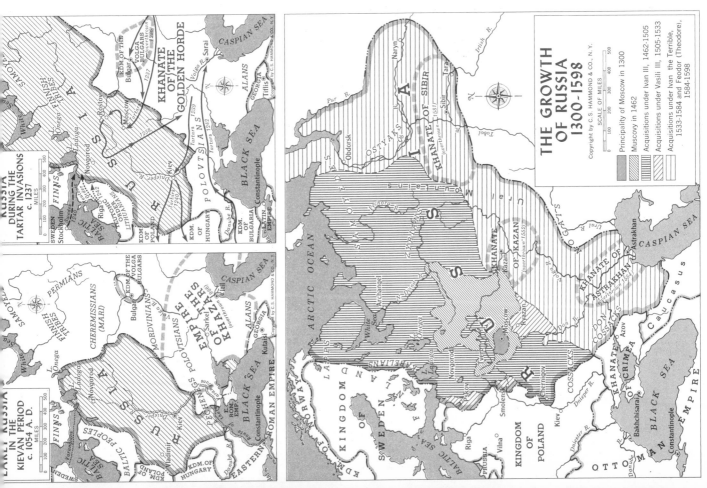

RUSSIA DURING THE TARTAR INVASIONS c. 1237

MILES

EARLY RUSSIA IN THE KIEVAN PERIOD c. 1054 A.D.

MILES

THE GROWTH OF RUSSIA 1300-1598

Copyright by C.S. HAMMOND & CO., N.Y.

SCALE OF MILES

	Principality of Moscow in 1300
	Muscovy in 1462
	Acquisitions under Ivan III, 1462-1505
	Acquisitions under Vasili III, 1505-1533
	Acquisitions under Ivan the Terrible, 1533-1584 and Feodor (Theodore), 1584-1598

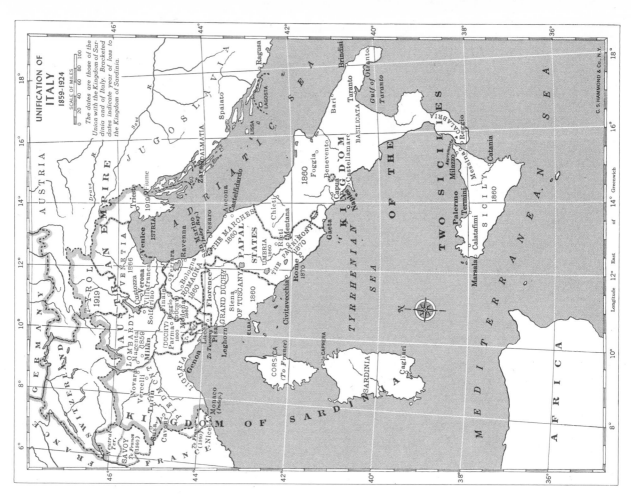

UNIFICATION OF
ITALY
1859-1924

SCALE OF MILES
0 20 40 60 80 100

The dates are those of the
Union with the Kingdom of Sardinia and of Italy. Bracketed
dates indicate year of loss to
the Kingdom of Sardinia.

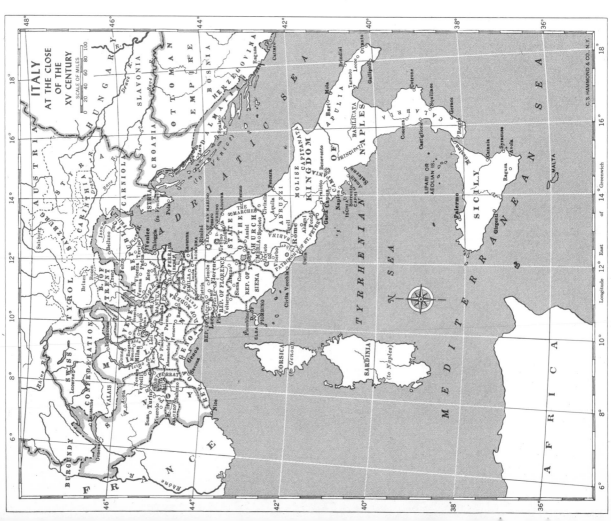

ITALY
AT THE CLOSE
OF THE
XV CENTURY

SCALE OF MILES
0 20 40 60 80 100

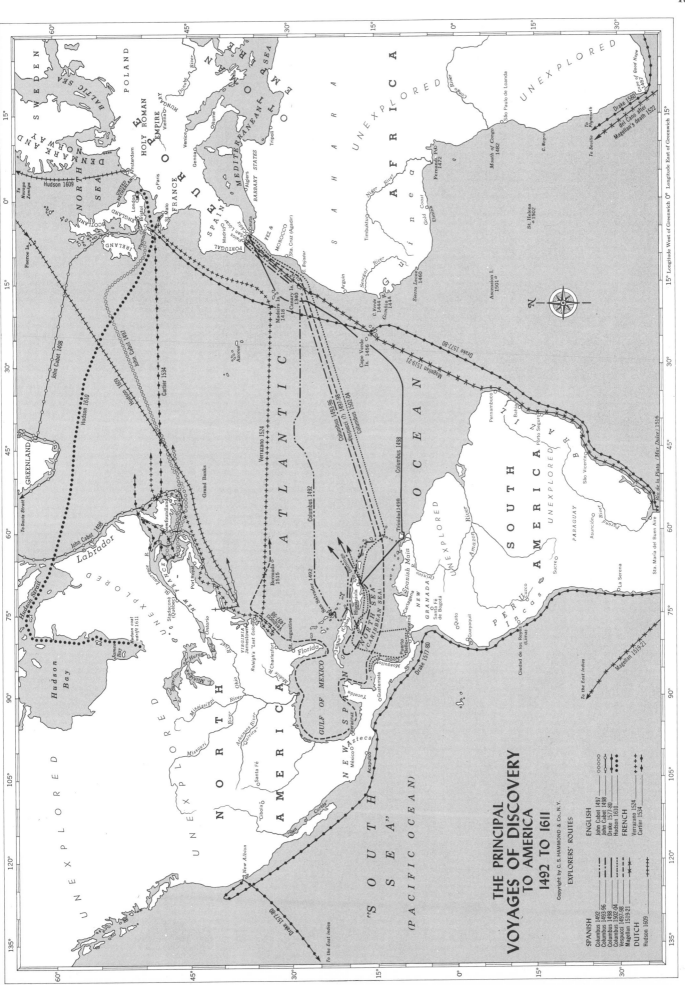

THE PRINCIPAL
VOYAGES OF DISCOVERY
TO AMERICA
1492 TO 1611

Copyright by C. S. HAMMOND & Co., N.Y.

EXPLORERS' ROUTES

SPANISH
Columbus 1492
Columbus 1493-96
Columbus 1498
Columbus 1502-04
Vespucci 1497-98
Magellan 1519-21
DUTCH
Hudson 1609

ENGLISH
John Cabot 1497
John Cabot 1498
Drake 1577-80
Hudson 1610
FRENCH
Verrazano 1524
Cartier 1534

UNITED STATES
Copyright by C. S. HAMMOND & Co., N. Y.

INDIAN TRIBES AND LINGUISTIC FAMILIES
after Powell and Kroeber

The various Indian tribes are shown where they were located during the period of their greatest significance in American history.

LINGUISTIC FAMILIES

1	ALGONQUIAN	12	KITUNAHAN	21	TONKAWAN	
2	ARAWAKAN	13	MUSKOGEAN	22	TUNICAN	
3	ATHAPASCAN	14	PIMAN	23	UCHEAN	
4	ATTACAPAN	15	SALISHAN	24	WAIILATPUAN	
5	CADDOAN	16	SHAHAPTIAN	25	WAKASHAN	
6	CHITIMACHAN	17	SHOSHONEAN	26	YUMAN	
7	COAHUILTECAN	18	SIOUAN	27	ZUÑIAN	
8	IROQUOIAN	19	TANOAN	28	Miscellaneous Pacific Coast Linguistic Families.	
9	KARANKAWAN	20	TIMUCUAN	29	Linguistic affinity uncertain.	
10	KERESAN					
11	KIOWAN					

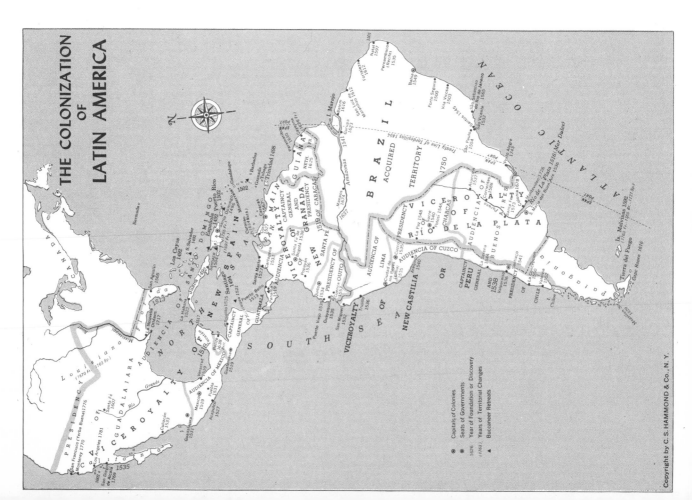

THE COLONIZATION OF LATIN AMERICA

Copyright by C. S. HAMMOND & Co., N. Y.

⊕ Capitals of Colonies
⊕ Seats of Governments
1526 Year of Foundation or Discovery
(1763) Years of Territorial Changes
▲ Buccaneer Retreats

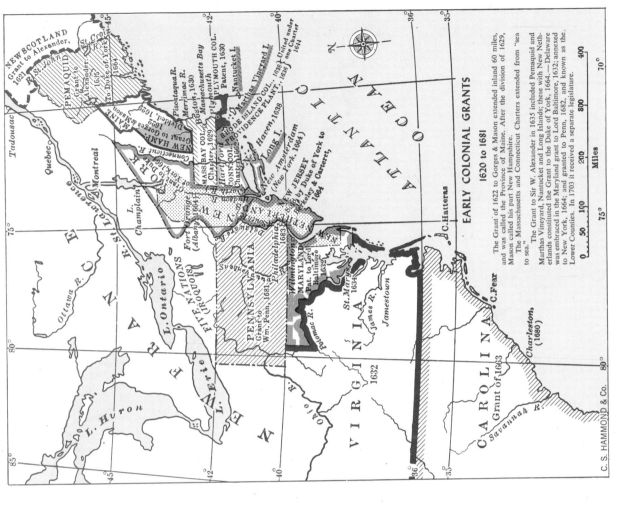

EARLY COLONIAL GRANTS
1620 to 1681

The Grant of 1622 to Gorges & Mason extended inland 60 miles, and was called the Province of Maine. After the division of 1629, Mason called his part New Hampshire.

The Massachusetts and Connecticut Charters extended from "sea to sea."

The Grant to Sir W. Alexander in 1635 included Pemaquid and Marthas Vineyard, Nantucket and Long Islands; these with New Netherlands constituted the Grant to the Duke of York, 1664.—Delaware was embraced in the Maryland grant to Lord Baltimore, 1632; annexed to New York, 1664; and granted to Penn, 1682, and known as the Lower Counties. In 1703 it received a separate legislature.

C. S. HAMMOND & CO.

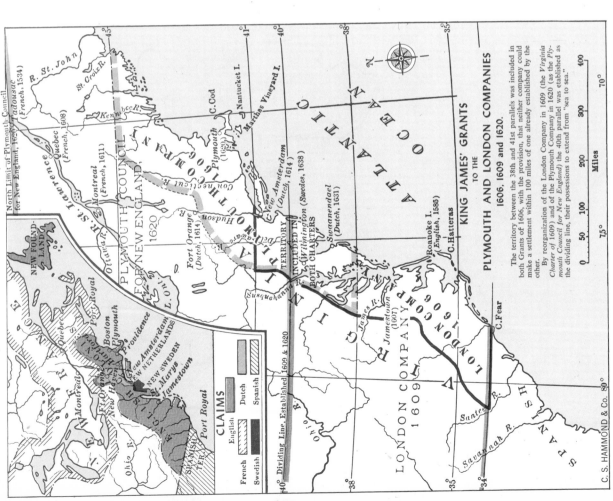

KING JAMES' GRANTS
TO THE
PLYMOUTH AND LONDON COMPANIES
1606, 1609 and 1620.

The territory between the 38th and 41st parallels was included in both Grants of 1606, with the provision, that neither company could make a settlement within 100 miles of one already established by the other.

By reorganization of the London Company in 1609 (the *Virginia Charter of 1609*) and of the Plymouth Company in 1620 (as the *Plymouth Council for New England*) the 40th parallel was established as the dividing line, their possessions to extend from "sea to sea."

CLAIMS: English, Dutch, Spanish, Swedish

C. S. HAMMOND & CO.

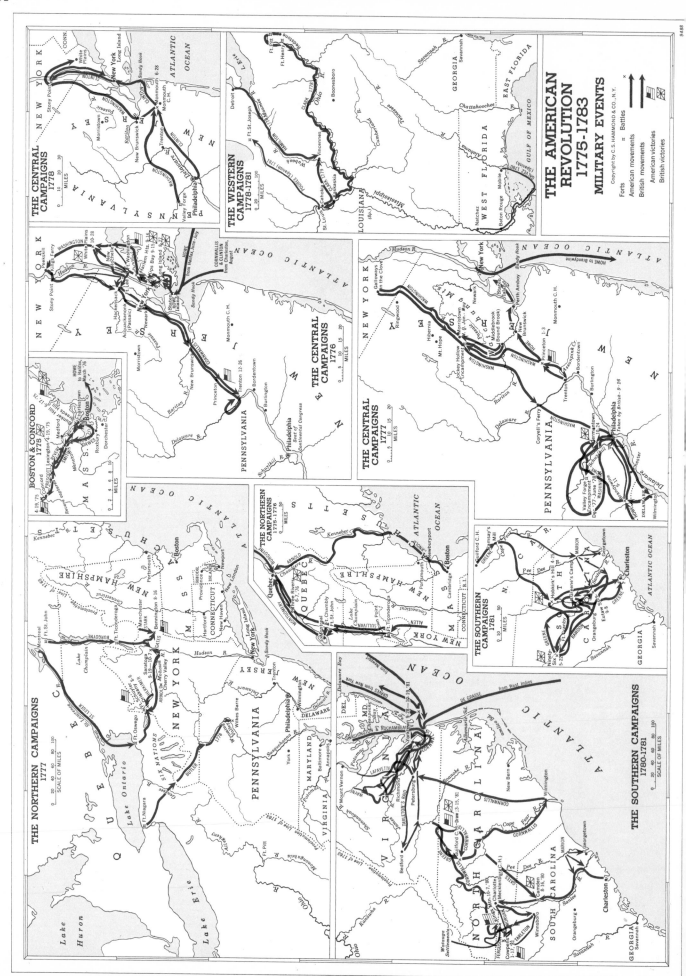

THE AMERICAN REVOLUTION 1775-1783 MILITARY EVENTS

Copyright by C.S. HAMMOND & CO., N.Y.

Forts ✕ Battles
→ American movements
→ British movements
⚑ American victories
⚑ British victories

THE CENTRAL CAMPAIGNS 1778

THE WESTERN CAMPAIGNS 1778-1781

THE CENTRAL CAMPAIGNS 1776

THE CENTRAL CAMPAIGNS 1777

BOSTON & CONCORD 1775

THE NORTHERN CAMPAIGNS 1775-1776

THE SOUTHERN CAMPAIGNS 1781

THE NORTHERN CAMPAIGNS 1777

THE SOUTHERN CAMPAIGNS 1780-1781

POLAND
TO 1667
Boundary of Poland previous to 1629
Lands ceded to Sweden in 1629 (confirmed 1660)
Lands ceded to Russia at the Peace of Andrussof, 1667

SCALE OF MILES
0   50 100   200   300

POLAND
RESULT OF THE
FIRST PARTITION, 1772
Boundary of Poland previous to 1772
Lands acquired by Russia
Lands acquired by Prussia
Lands acquired by Austria

SCALE OF MILES
0   50 100   200   300

POLAND
RESULT OF THE
SECOND PARTITION, 1793
Boundary of Poland from 1772 to 1793
Lands acquired by Russia
Lands acquired by Prussia
Austria took no part in this partition.

SCALE OF MILES
0   50 100   200   300

POLAND
RESULT OF THE
THIRD PARTITION, 1795
Boundary of Poland from 1793 to 1795
Lands acquired by Russia
Lands acquired by Prussia
Lands acquired by Austria

SCALE OF MILES
0   50 100   200   300

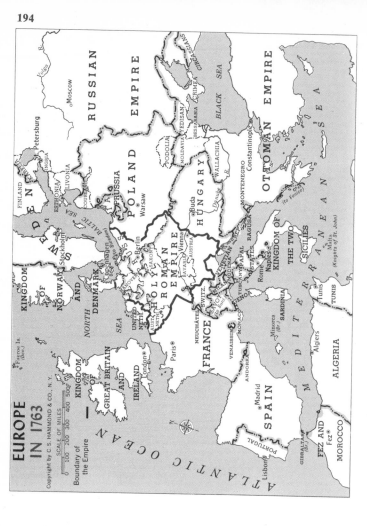

EUROPE IN 1648

Copyright by C. S. HAMMOND & CO., N.Y.

SCALE OF MILES
0 100 200 300 400

Boundary of the Empire

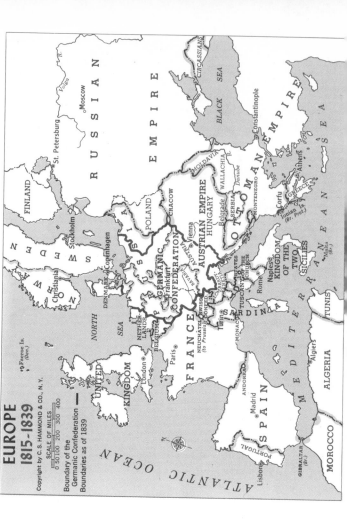

EUROPE IN 1763

Copyright by C. S. HAMMOND & CO., N.Y.

SCALE OF MILES
0 100 200 300 400 500

Boundary of the Empire

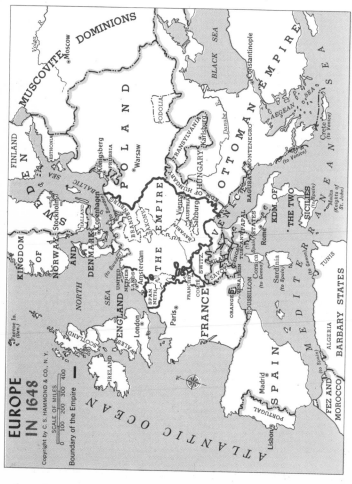

EUROPE IN 1812

Copyright by C. S. HAMMOND & CO., N.Y.

SCALE OF MILES
0 100 200 300 400 500

Boundary of the Confederation of the Rhine

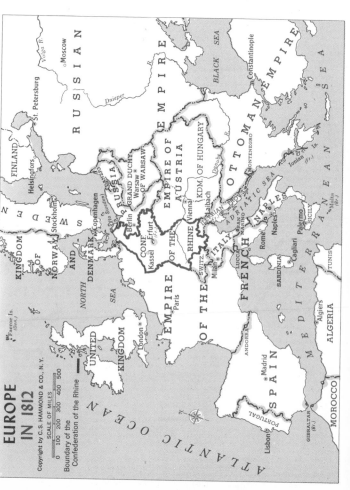

EUROPE 1815-1839

Copyright by C. S. HAMMOND & CO., N.Y.

SCALE OF MILES
0 50 100 200 300 400

Boundary of the Germanic Confederation
Boundaries as of 1839

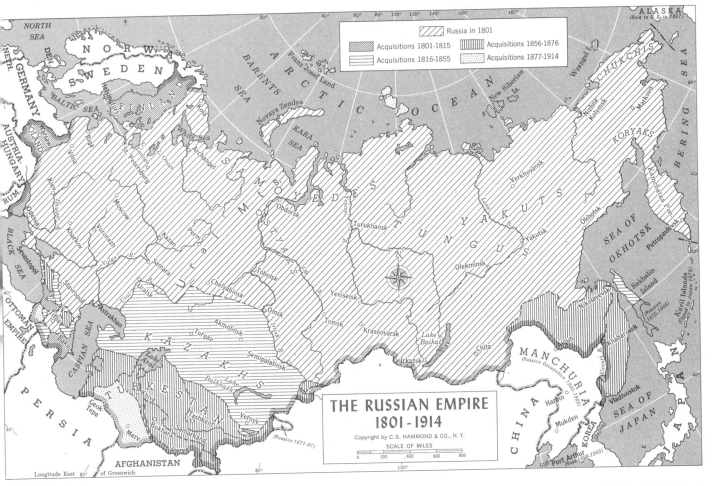

THE RUSSIAN EMPIRE
1801-1914

Copyright by C.S. HAMMOND & CO., N.Y.

SCALE OF MILES

| Russia in 1801 |
| Acquisitions 1801-1815 | Acquisitions 1856-1876 |
| Acquisitions 1816-1855 | Acquisitions 1877-1914 |

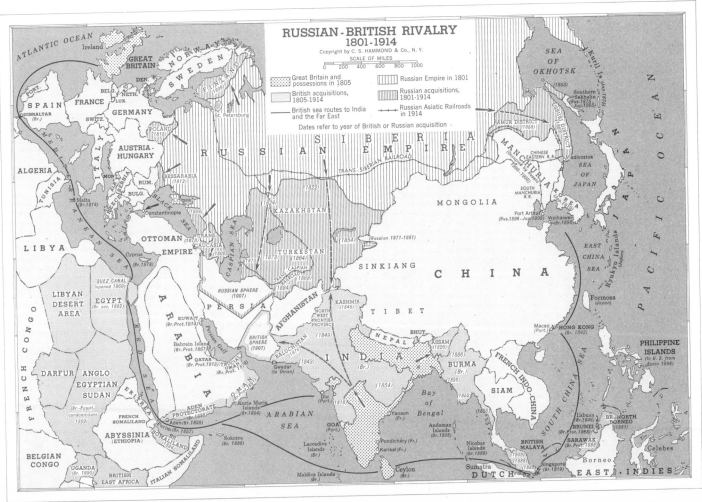

RUSSIAN-BRITISH RIVALRY
1801-1914

Copyright by C. S. HAMMOND & Co., N.Y.

SCALE OF MILES

Great Britain and possessions in 1805 — Russian Empire in 1801
British acquisitions, 1805-1914 — Russian acquisitions, 1801-1914
British sea routes to India and the Far East — Russian Asiatic Railroads in 1914

Dates refer to year of British or Russian acquisition

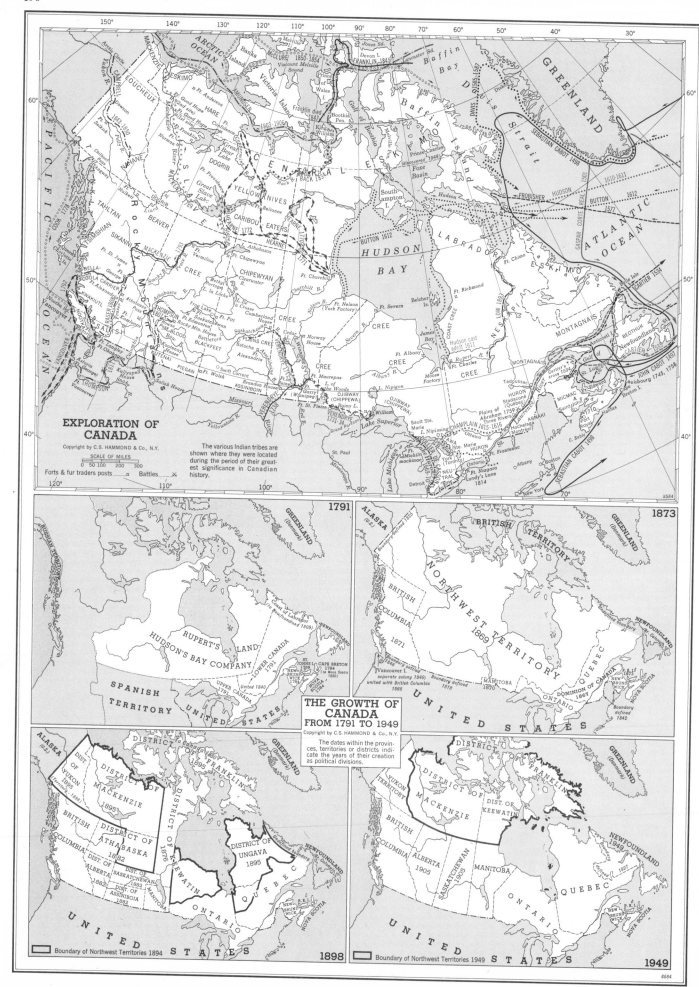

EXPLORATION OF CANADA

Copyright by C.S. Hammond & Co., N.Y.

SCALE OF MILES
0 50 100 200 300

Forts & fur traders posts ●   Battles ✕

The various Indian tribes are shown where they were located during the period of their greatest significance in Canadian history.

THE GROWTH OF CANADA
FROM 1791 TO 1949

Copyright by C.S. Hammond & Co., N.Y.

The dates within the provinces, territories or districts indicate the years of their creation as political divisions.

1791

1873

Boundary of Northwest Territories 1894   1898

Boundary of Northwest Territories 1949   1949

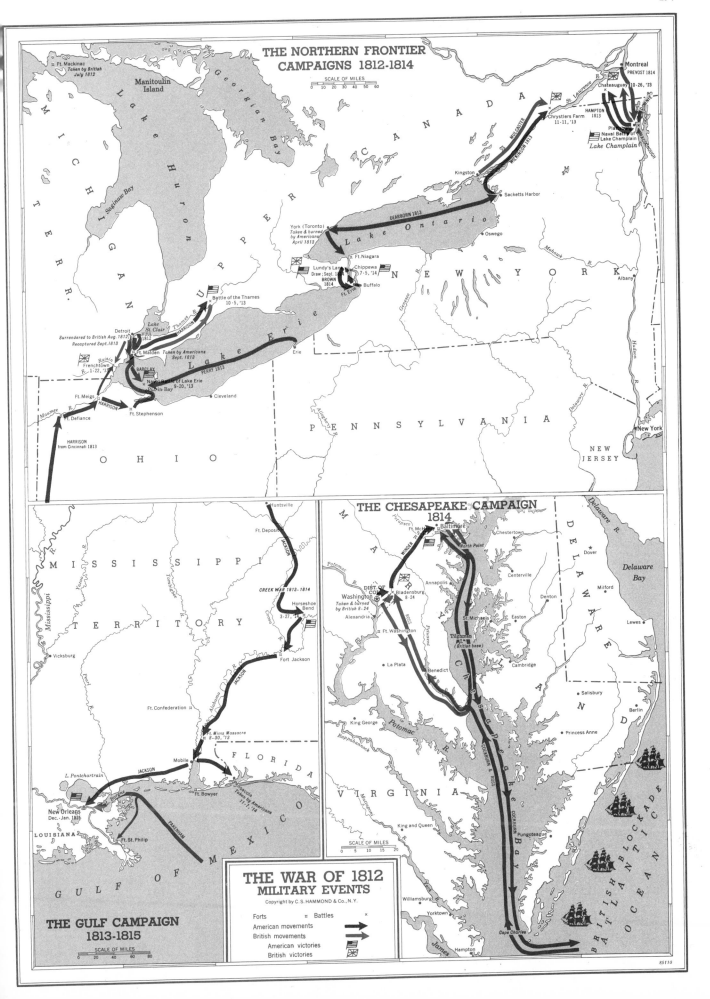

THE NORTHERN FRONTIER
CAMPAIGNS 1812-1814

SCALE OF MILES
0 10 20 30 40 50 60

U P P E R   C A N A D A

Ft. Mackinac
Taken by British July 1812

Manitoulin Island

Lake Huron

Georgian Bay

M I C H I G A N   T E R R.

Saginaw Bay

Lake St. Clair

Detroit
HULL 1812
Surrendered to British Aug.1812. Recaptured Sept.1813

Ft. Malden *Taken by Americans Sept. 1813*

Raisin R.
Frenchtown
1-22, '13

Battle of the Thames
10-5, '13

Thames R.
HARRISON

Ft. Meigs
HARRISON

Ft. Defiance

Ft. Stephenson

HARRISON from Cincinnati 1813

Maumee R.

O H I O

BARCLAY
Naval Battle of Lake Erie
Put-in-Bay 9-20, '13
PERRY 1813

Lake Erie

Cleveland

Erie

York (Toronto)
Taken & burned by Americans April 1813

Lake Ontario

DEARBORN 1813

Kingston

Sacketts Harbor

Oswego

Ft. Niagara
Lundy's Lane Draw; Sept. 1814
Chippewa 7-5, '14
BROWN 1814
Ft. Erie
Buffalo

Genesee R.

Mohawk R.

Albany

N E W   Y O R K

WILKINSON 1813
MULCASTER 1813
Chrystlers Farm 11-11, '13

St. Lawrence R.

Montreal
PREVOST 1814

Chateauguay 10-26, '23
HAMPTON 1813
Plattsburg
Naval Battle of Lake Champlain
Lake Champlain
MACDONOUGH

P E N N S Y L V A N I A

Allegheny R.

Susquehanna R.

Delaware R.

Hudson R.

New York

N E W   J E R S E Y

THE GULF CAMPAIGN
1813-1815

SCALE OF MILES
0 20 40 60 80

M I S S I S S I P P I   T E R R I T O R Y

Mississippi R.

Yazoo R.

Vicksburg

Pearl R.

Tombigbee R.

Alabama R.

Huntsville

Ft. Deposit
JACKSON

CREEK WAR 1813-1814

Horseshoe Bend
3-27, '14

Coosa R.

Ft. Jackson

Ft. Confederation

Ft. Mims Massacre
8-30, '13

JACKSON

Mobile

FLORIDA

L. Pontchartrain

New Orleans
Dec.-Jan. 1815

L. Borgne

JACKSON

Ft. Bowyer

Pensacola *Taken by Americans 11-7, '14*

L O U I S I A N A

Ft. St. Philip

PAKENHAM

G U L F   O F   M E X I C O

THE CHESAPEAKE CAMPAIGN
1814

M A R Y L A N D

Patapsco R.

Ft. McHenry
Baltimore

North Point

WINDER

Potomac R.

DIST. OF COL.
Washington *Taken & burned by British 8-24*
Bladensburg 8-24
ROSS
Alexandria
Ft. Washington

Annapolis

Chestertown

D E L A W A R E

Dover

Centerville

Delaware Bay

Milford

Denton

Lewes

Easton

St. Michaels

Tilghman I.
(British base)

Patuxent R.

La Plata

Benedict

Cambridge

Salisbury

Berlin

Princess Anne

M A R Y L A N D

King George

Rappahannock R.

V I R G I N I A

Mattaponi R.

King and Queen

Pungoteague

COCKBURN; ROSS

Chesapeake Bay

COCKBURN

Pamunkey R.

Williamsburg

Yorktown

York R.

James R.

Hampton

Cape Charles

B R I T I S H   B L O C K A D E

A T L A N T I C   O C E A N

SCALE OF MILES
0 5 10 15 20

THE WAR OF 1812
MILITARY EVENTS
Copyright by C.S. Hammond & Co., N.Y.

Forts ⊔	Battles ×
American movements →	
British movements ⇒	
American victories 🏴	
British victories 🏴	

85110

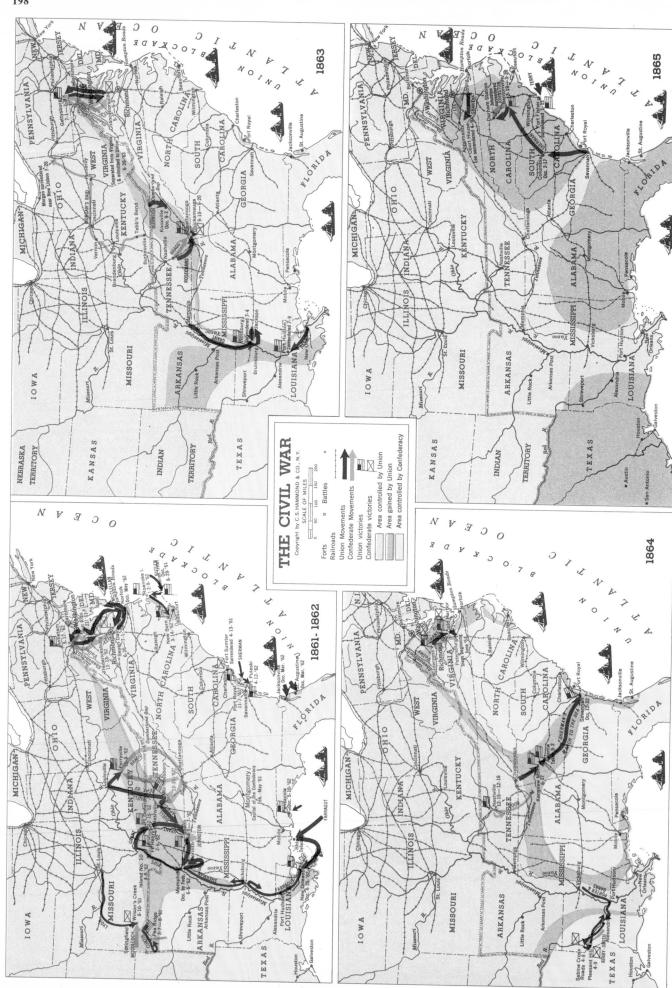

THE CIVIL WAR

Copyright by C.S. HAMMOND & CO., N.Y.

SCALE OF MILES

0 50 100 150 200

Forts

Railroads

Union Movements

Confederate Movements

Union victories

Confederate victories

Battles

Area controlled by Union

Area gained by Union

Area controlled by Confederacy

1861-1862

1863

1864

1865

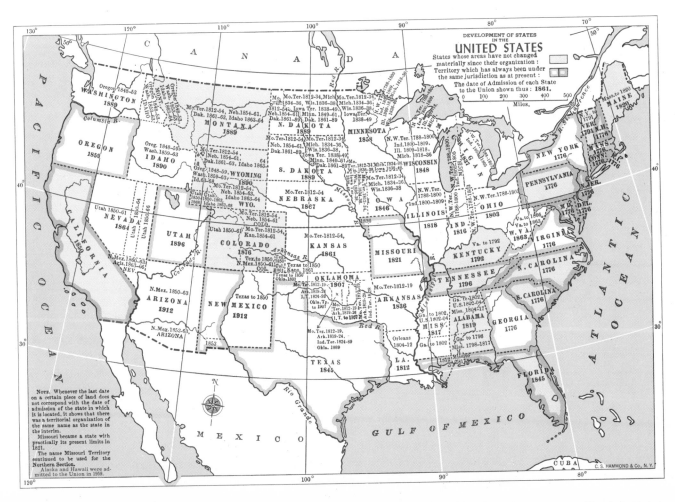

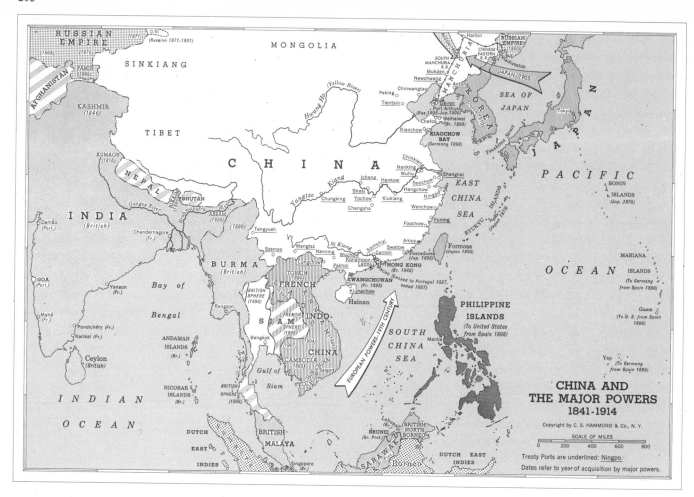

CHINA AND THE MAJOR POWERS
1841-1914

Copyright by C. S. HAMMOND & Co., N. Y.

SCALE OF MILES

0   200   400   600   800

Treaty Ports are underlined: Ningpo.

Dates refer to year of acquisition by major powers.

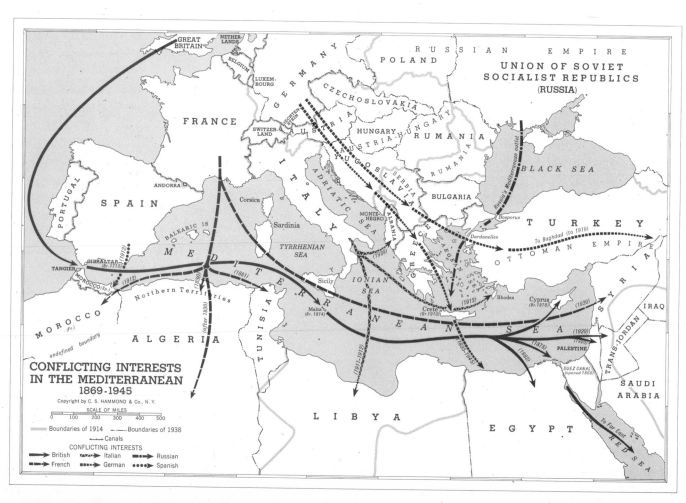

CONFLICTING INTERESTS IN THE MEDITERRANEAN
1869-1945

Copyright by C. S. HAMMOND & Co., N. Y.

SCALE OF MILES

0   100   200   300   400   500

—————— Boundaries of 1914   —— —— Boundaries of 1938

———·——— Canals

CONFLICTING INTERESTS

→ British   ◁◁◁ Italian   ▷▷▷ Russian

–·–· French   ▪▪▪ German   •••• Spanish

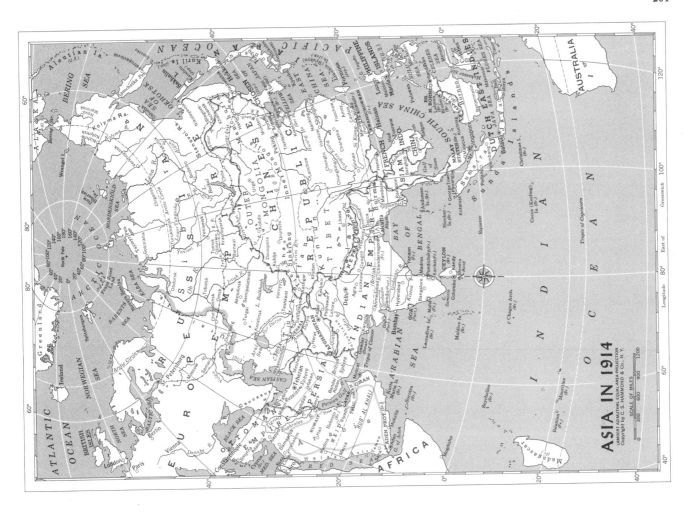

ASIA IN 1914

LAMBERT AZIMUTHAL EQUAL-AREA PROJECTION
Copyright by C. S. HAMMOND & Co., N.Y.

SCALE OF MILES
0 300 600 900 1200

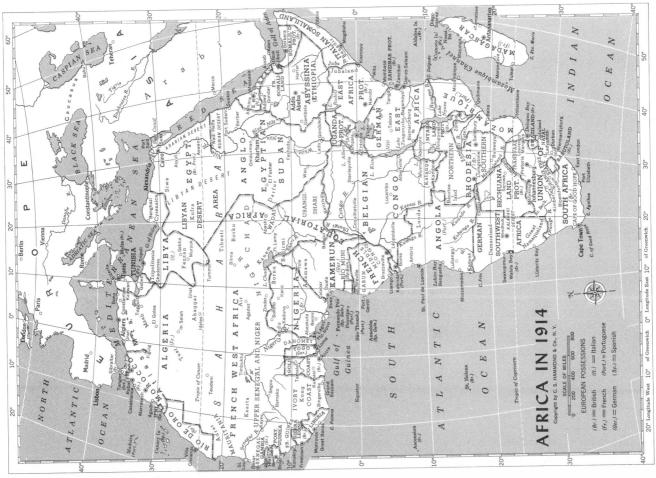

AFRICA IN 1914

Copyright by C. S. HAMMOND & Co., N.Y.

SCALE OF MILES
0 200 400 600 800

EUROPEAN POSSESSIONS

(Br.) = British   (It.) = Italian
(Fr.) = French   (Port.) = Portuguese
(Ger.) = German   (Sp.) = Spanish

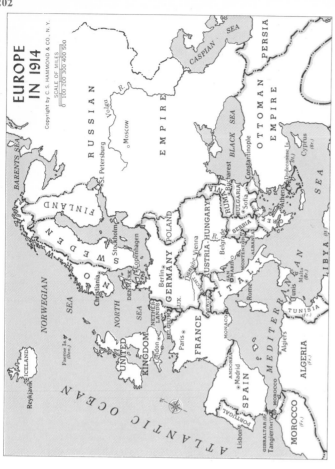

EUROPE IN 1914

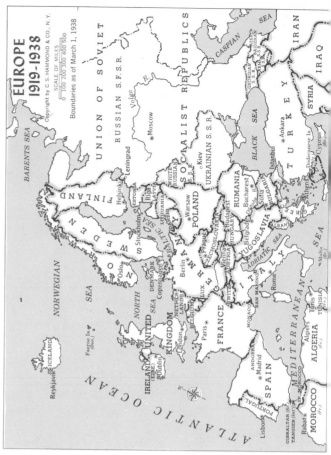

EUROPE 1919-1938

Boundaries as of March 1, 1938

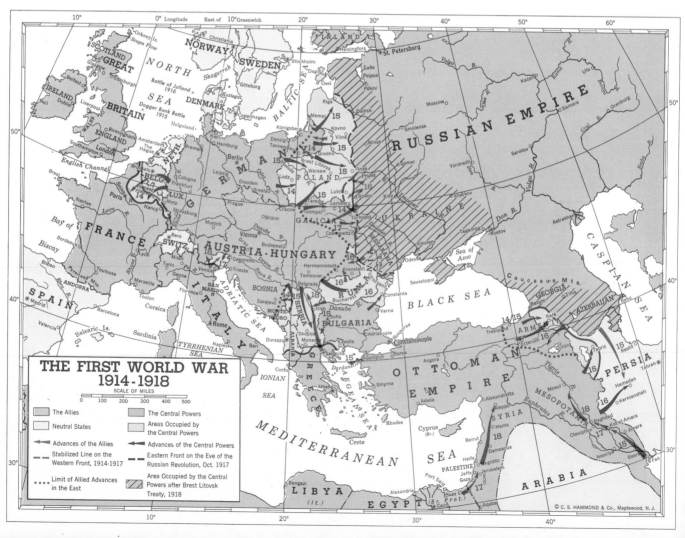

THE FIRST WORLD WAR
1914-1918

SCALE OF MILES

0   100   200   300   400   500

▨ The Allies	▨ The Central Powers	
☐ Neutral States	▨ Areas Occupied by the Central Powers	

← Advances of the Allies ← Advances of the Central Powers

– – Stabilized Line on the Western Front, 1914-1917

― Eastern Front on the Eve of the Russian Revolution, Oct. 1917

···· Limit of Allied Advances in the East

▨ Area Occupied by the Central Powers after Brest Litovsk Treaty, 1918

© C. S. HAMMOND & Co., Maplewood, N.J.

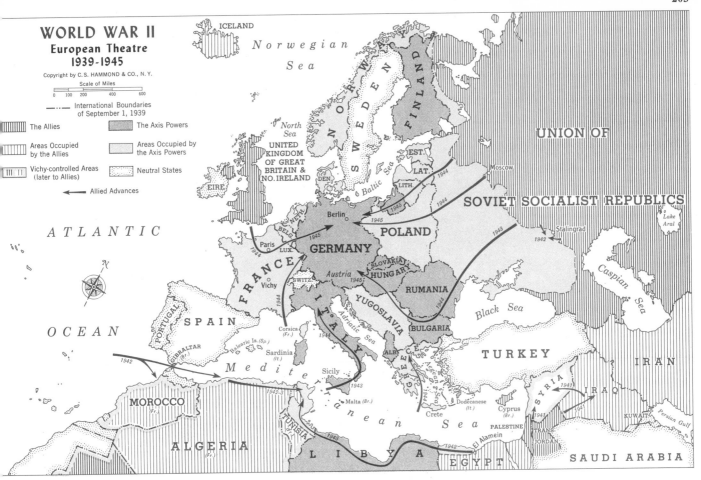

WORLD WAR II
European Theatre
1939-1945

Copyright by C.S. HAMMOND & CO., N.Y.

Scale of Miles
0   100   200   400   600

----- International Boundaries
of September 1, 1939

|||| The Allies

||| Areas Occupied
by the Allies

|||| Vichy-controlled Areas
(later to Allies)

The Axis Powers

Areas Occupied by
the Axis Powers

Neutral States

← Allied Advances

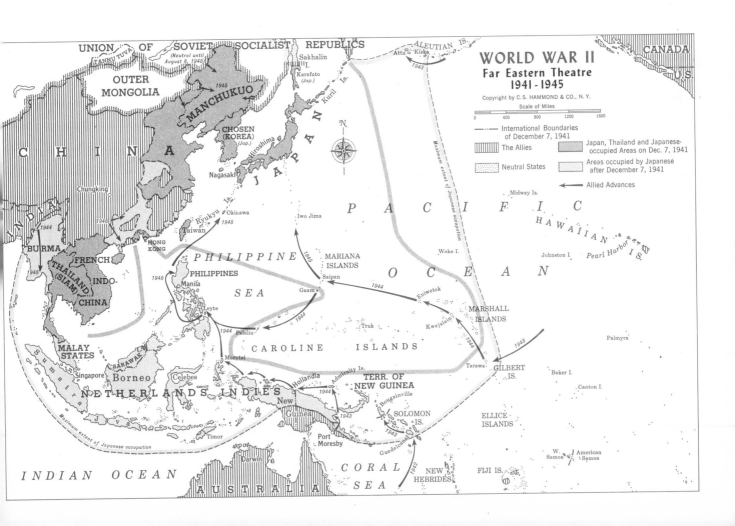

WORLD WAR II
Far Eastern Theatre
1941-1945

Copyright by C.S. HAMMOND & CO., N.Y.

Scale of Miles
0   400   800   1200   1600

----- International Boundaries
of December 7, 1941

|||| The Allies

Neutral States

Japan, Thailand and Japanese-
occupied Areas on Dec. 7, 1941

Areas occupied by Japanese
after December 7, 1941

← Allied Advances

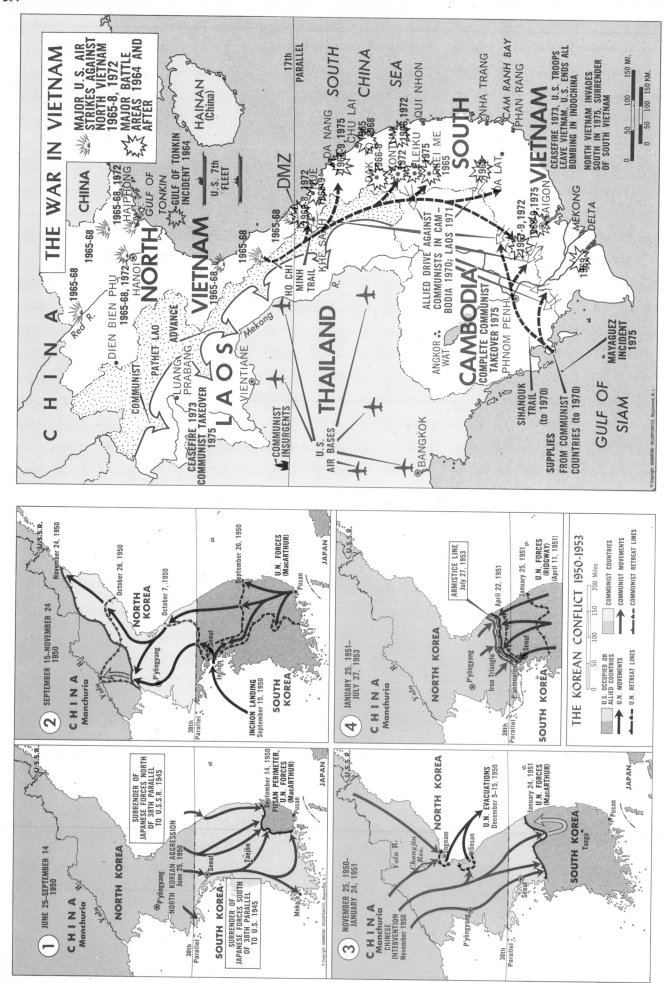

THE WAR IN VIETNAM

MAJOR U.S. AIR STRIKES AGAINST NORTH VIETNAM 1965-8, 1972

MAJOR BATTLE AREAS 1964 AND AFTER

GULF OF TONKIN INCIDENT 1964

U.S. 7th FLEET

CHINA

HAINAN (China)

17th PARALLEL

SOUTH CHINA SEA

NORTH VIETNAM

DMZ

DIEN BIEN PHU 1965-68, 1972

HANOI

HAIPHONG 1965-68, 1972

GULF OF TONKIN

Red R.

PATHET LAO

COMMUNIST ADVANCE

LUANG PRABANG

LAOS

VIENTIANE

Mekong

HO CHI MINH TRAIL

KHE SANH

HUE 1968-8, 1972 1969

DA NANG 1965-9, 1975

CHU LAI 1965

KONTUM 1972 1972 1966-8 1975

PLEIKU 1975

DAK TO 1966-8 1968

MY LAI 1968

BAN ME THUOT 1975

QUI NHON

NHA TRANG

CAM RANH BAY

PHAN RANG

DA LAT 1965

THAILAND

U.S. AIR BASES

ALLIED DRIVE AGAINST COMMUNISTS IN CAMBODIA 1970; LAOS 1971

ANGKOR WAT

CAMBODIA

COMPLETE COMMUNIST TAKEOVER 1975

PHNOM PENH

SAIGON 1965, 1975

1965-9, 1972

MEKONG DELTA

1969

SOUTH VIETNAM

SIHANOUK TRAIL (to 1970)

SUPPLIES FROM COMMUNIST COUNTRIES (to 1970)

GULF OF SIAM

MAYAGUEZ INCIDENT 1975

BANGKOK

COMMUNIST INSURGENTS

CEASEFIRE 1973, U.S. TROOPS LEAVE VIETNAM, U.S. ENDS ALL BOMBING IN INDOCHINA

NORTH VIETNAM INVADES SOUTH IN 1975, SURRENDER OF SOUTH VIETNAM

CEASEFIRE 1973 COMMUNIST TAKEOVER 1975

150 MI.
150 KM.

© Copyright HAMMOND INCORPORATED, Maplewood, N.J.

---

① JUNE 25–SEPTEMBER 14 1950

U.S.S.R.

CHINA
Manchuria

Yalu

NORTH KOREA

Pyongyang

Seoul

SURRENDER OF JAPANESE FORCES NORTH OF 38TH PARALLEL TO U.S.S.R. 1945

September 14, 1950

PUSAN PERIMETER, U.N. FORCES (MacARTHUR)
Pusan

NORTH KOREAN AGGRESSION June 25, 1950

Taejon

SOUTH KOREA

Mokpo

SURRENDER OF JAPANESE FORCES SOUTH OF 38TH PARALLEL TO U.S. 1945

38th Parallel

JAPAN

---

② SEPTEMBER 15–NOVEMBER 24 1950

U.S.S.R.

November 24, 1950

October 26, 1950

CHINA
Manchuria

Yalu

October 7, 1950

NORTH KOREA

Pyongyang

September 26, 1950

U.N. FORCES (MacARTHUR)
Pusan

Seoul

Inchon

INCHON LANDING September 15, 1950

SOUTH KOREA

38th Parallel

JAPAN

---

③ NOVEMBER 25, 1950– JANUARY 24, 1951

U.S.S.R.

CHINA
Manchuria

CHINESE INTERVENTION November 1950

Yalu R.

Changjin Res.

Hungnam

Wonsan

U.N. EVACUATIONS December 5–15, 1950

NORTH KOREA

Pyongyang

Seoul

January 24, 1951 U.N. FORCES (MacARTHUR)
Pusan

SOUTH KOREA

Taegu

38th Parallel

JAPAN

© Copyright HAMMOND INCORPORATED, Maplewood, N.J.

---

④ JANUARY 25, 1951– JULY 27, 1953

U.S.S.R.

CHINA
Manchuria

Yalu

NORTH KOREA

Pyongyang

ARMISTICE LINE July 27, 1953

Iron Triangle

Panmunjom

April 22, 1951

January 25, 1951

U.N. FORCES (RIDGWAY) (April 11, 1951)

Seoul

SOUTH KOREA

38th Parallel

JAPAN

THE KOREAN CONFLICT 1950-1953

0 50 100 150 200 Miles

U.S. OCCUPIED OR ALLIED COUNTRIES

U.N. MOVEMENTS

U.N. RETREAT LINES

COMMUNIST COUNTRIES

COMMUNIST MOVEMENTS

COMMUNIST RETREAT LINES

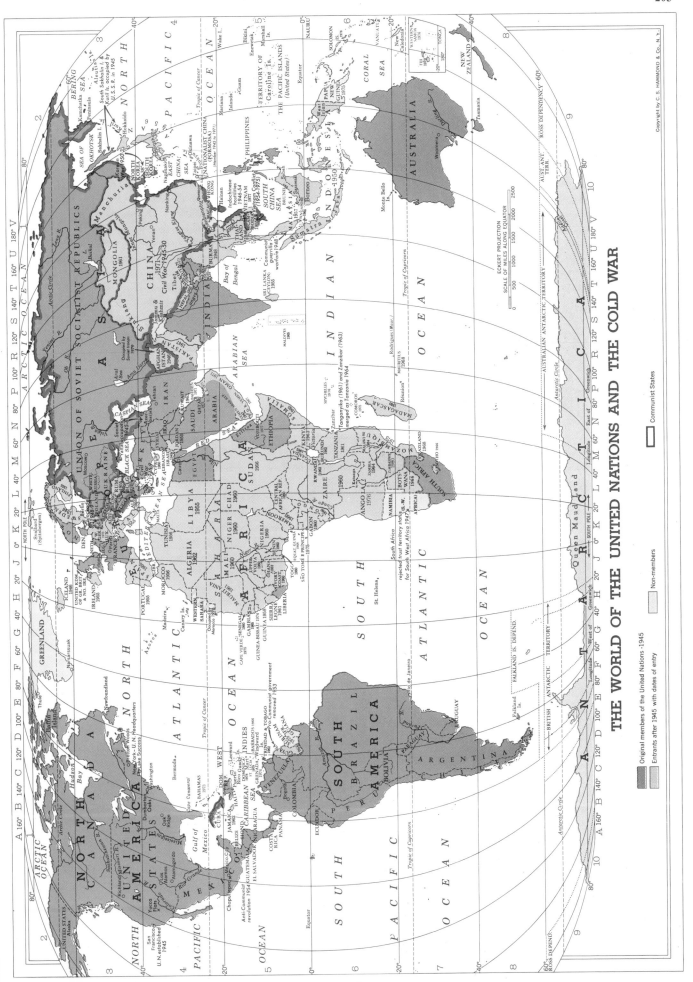

THE WORLD OF THE UNITED NATIONS AND THE COLD WAR

Original members of the United Nations - 1945

Entrants after 1945 with dates of entry

Non-members

Communist States

ECKERT PROJECTION
SCALE OF MILES ALONG EQUATOR

THE WORLD TODAY

Western Hemisphere

Copyright by C. S. HAMMOND & Co., N.Y.

SCALE OF MILES
0 500 1000 1500 20

SCALE OF KILOMETRES
0 500 1000 1500 2000

Capitals of Countries ....

International Boundaries ___ — ___

THE UNITED NATIONS

Afghanistan	Brazil	Costa Rica								Tunisia
Albania	Bulgaria	Cuba	Gabon	India	Lesotho	Mozambique	Philippines	Somalia	Turkey
Algeria	Burma	Cyprus	Gambia	Indonesia	Liberia	Nepal	Poland	South Africa	Uganda
Angola	Burundi	Czechoslovakia	Germany, East	Iran	Libya	Netherlands	Portugal	Soviet Union	United Arab Emirates
Argentina	Cambodia	Denmark	Germany, West	Iraq	Luxembourg	New Zealand	Qatar	Byelorussian S.S.R.	United Kingdom
Australia	Cameroon	Djibouti	Ghana	Ireland	Madagascar	Nicaragua	Rumania	Ukrainian S.S.R.	United States
Austria	Canada	Dominica	Greece	Israel	Malawi	Niger	Rwanda	Spain	Upper Volta
Bahamas	Cape Verde	Dominican Rep.	Grenada	Italy	Malaysia	Nigeria	St. Lucia	Sri Lanka	Uruguay
Bahrain	Central African	Ecuador	Guatemala	Ivory Coast	Maldives	Norway	São Tomé e	Sudan	Venezuela
Bangladesh	Republic	Egypt	Guinea	Jamaica	Mali	Oman	Príncipe	Suriname	Vietnam
Barbados	Chad	El Salvador	Guinea-Bissau	Japan	Malta	Pakistan	Saudi Arabia	Swaziland	Western Samoa
Belgium	Chile	Equatorial Guinea	Guyana	Jordan	Mauritania	Panama	Senegal	Sweden	Yemen, P.D.R. of
Benin	China	Ethiopia	Haiti	Kenya	Mauritius	Papua	Seychelles	Syria	Yemen Arab Rep.
Bhutan	Colombia	Fiji	Honduras	Kuwait	Mexico	New Guinea	Sierra Leone	Tanzania	Yugoslavia
Bolivia	Comoros	Finland	Hungary	Laos	Mongolia	Paraguay	Singapore	Thailand	Zaire
Botswana	Congo	France	Iceland	Lebanon	Morocco	Peru	Solomon Islands	Togo	Zambia
								Trinidad and Tobago	Zimbabwe

THE WORLD TODAY
Eastern Hemisphere

Copyright by C.S. Hammond & Co., N.Y.

SCALE OF MILES

500   1000   1500   2000

SCALE OF KILOMETRES

0   500   1000 1500 2000

Capitals of Countries ....◎

International Boundaries ......_____

NORTH ATLANTIC TREATY ORGANIZATION (NATO)

Belgium

Canada

Denmark

France

German Federal

  Rep. (West Ger.)

Greece

Iceland

Italy

Luxembourg

Netherlands

Norway

Portugal

Turkey

United Kingdom

United States

ORGANIZATION OF AMERICAN STATES (OAS-Rio Pact)

Argentina

Barbados

Bolivia

Brazil

Chile

Colombia

Costa Rica

Dominica

Dominican Rep.

Ecuador

El Salvador

Grenada

Guatemala

Haiti

Honduras

Jamaica

Mexico

Nicaragua

Panama

Paraguay

Peru

St. Lucia

Trinidad/Tob.

United States

Uruguay

Venezuela

Cuba *

* Expelled from activities of O.A.S.

ASSOCIATION OF SOUTHEAST ASIAN NATIONS (ASEAN)

Indonesia

Malaysia

Philippines

Singapore

Thailand

WARSAW PACT COUNTRIES ("SOVIET BLOC")

Soviet Union

Bulgaria

Czechoslovakia

German Democratic

  Rep. (East Ger.)

Hungary

Poland

Rumania

VOYAGES OF DISCOVERY & EXPLORATION

Year	Explorer and Nationality	Discovery — Exploration — Journey
B. C.		
500	Himilco (Carthage)	Explores Atlantic Coast of Europe.
470	Hanno (Carthage)	Leads a colonizing expedition to West Africa, as far as Cape Palmas.
330	Pytheas of Massilia (Greek)	Explores coast of Spain, Gaul and Great Britain.
332-326	Alexander the Great (Macedonia)	Enters India in 326.
325	Nearcus (Macedonia)	Sails from the Indus to the Euphrates River.
120	Eudoxus of Cnidus (Greek)	Attempts circumnavigation of Africa.
A. D.		
84	Gnaeus Julius Agricola (Roman)	Circumnavigates Great Britain.
861	Norsemen	Explore Faeroe Islands and Iceland; round north cape of Europe.
876	Gunnbjörn (Norse)	Sights Greenland Coast.
982	Eric the Red (Norse)	Discovers and names Greenland.
1000	Leif Ericson (Norse)	Discovers Newfoundland (Helluland), and Coast of New England (Vinland).
1160	Benjamin of Tudela (Navarre)	Travels in Turkey, Egypt, Assyria and Persia, penetrating to the frontiers of China.
1200	Arabs	Trading merchants discover Siberia.
1253	Jan van Ruysbroek (Dutch)	Reaches Karakorum, the ancient seat of the Mongol Empire.
1271-1295	Marco Polo (Venetian)	Travels in Central Asia, India, Persia. First to travel in China.
1325-1352	Ibn Batuta (Arabian)	Travels through North Africa, East Africa, South Russia, Arabia, India and China.
1487	Bartholomew Dias (Portuguese)	Rounds Cape of Good Hope to a point beyond Algoa Bay.
1492-1494	Christopher Columbus (Genoan)	Discovers the West Indies on Oct. 12, 1492; Dominica, Puerto Rico and several of the Windward Islands on Nov. 3, 1493 during his second voyage; Jamaica on May 3, 1494.
1497	Amerigo Vespucci (Florentine)	Discovers Venezuela and the continent of South America.
1497	John Cabot (Anglo-Venetian)	Sails along the northeast coast of America, discovering Cape Breton Islands and Nova Scotia.
1498	Vasco da Gama (Portuguese)	Takes route to India via Cape of Good Hope.
1498	Sebastian Cabot (English)	Explores American Coast from Gulf of Saint Lawrence to Chesapeake Bay.
1499	Christopher Columbus (Genoan)	Discovers Trinidad on July 31st; enters mouth of Orinoco River on August 1st.
1499	Alonso de Ojeda (Italian)	Discovers Gulf of Venezuela and New Granada.
1500	Vicente Pinzon (Spanish)	Discovers mouth of the Amazon.
1501	Pedro Alvarez Cabral (Portuguese)	Explores Coast of Brazil which he names Santa Cruz.
1502	Amerigo Vespucci (Florentine)	Discovers Bay of Rio de Janeiro.
1502	Christopher Columbus (Genoan)	Visits Central America on his fourth voyage; discovers Martinique.
1513	Ponce de Leon (Spanish)	Discovers Florida and sails up the west coast of the Peninsula.
1513	Vasco Nunez de Balboa (Spanish)	Crosses Isthmus of Panama and discovers the Pacific Ocean.
1518	Juan de Grijalva (Spanish)	Discovers east coast of Mexico.
1519-1521	Hernando Cortez (Spanish)	Conquers Mexico.
1519-1521	Ferdinand Magellan (Spanish), del Cano (after Magellan's death)	First to circumnavigate the globe. Passes thru the Strait of Magellan, crosses the Pacific and discovers the Philippines.
1524	Giovanni Verrazano (Italian)	Explores coast of North Carolina, Maryland, New Jersey and New York.
1524-1535	Jacques Cartier (French)	Explores Gulf of St. Lawrence and ascends river to Montreal.
1534	Francisco Pizarro (Spanish)	Completes the conquest of Peru.
1540-1541	Francisco de Orellana (Spanish)	Crosses Andes east of Quito and descends Amazon River to mouth.
1541	Fernando de Soto (Spanish)	Discovers the Mississippi River.
1576	Sir Martin Frobisher (English)	Explores Labrador and Baffin Bay; discovers Frobisher Bay.
1577-1580	Sir Francis Drake (English)	Second circumnavigation of the globe; explores west coast of North America as far as Oregon.
1592	John Davis (English)	Discovers the Falkland Islands.

Year	Explorer and Nationality	Discovery — Exploration — Journey
1595	Sir Walter Raleigh (English)	Explores Guiana and ascends the Orinoco 4__ miles.
1596	William Barents (Dutch)	Discovers Spitzbergen, Nova Zemlya a__ Barents Sea.
1606	Pedro Fernandez de Queiros (Spanish)	Discovers the New Hebrides.
1608	Samuel de Champlain (French)	Discovers Lake Ontario.
1608	John Smith (English)	Explores Chesapeake Bay and its tributaries.
1609-1610	Henry Hudson (English)	Explores Hudson River and Hudson Bay.
1616	William Baffin (English)	Enters Baffin Bay in quest of Northwe__ Passage.
1642	Abel Tasman (Dutch)	Discovers Van Dieman's Land (Tasmania) a__ New Zealand.
1673	Jacques Marquette and Louis Joliet (French)	Explore the Mississippi River from the nort__
1681	Renè R. C. La Salle (French)	Explores Lower Mississippi and takes possessio__ for Louis XIV.
1701	Father Kino	Explores California.
1728	Vitus Bering (Dane)	Discovers Bering Strait and proves that As__ and America are not connected.
1768-1775	Capt. James Cook (English)	Circumnavigates the globe and makes hydro__ graphic surveys of the Society Islands, San__ wich Islands, east coast of Australia, Coo__ Strait in New Zealand.
1789-1793	Alex. Mackenzie (Scotch)	Explores Mackenzie River; first to make over__ land trip to Pacific Coast.
1792	George Vancouver (English)	Circumnavigates Vancouver Island, explor__ Gulf of Georgia.
1803-1806	Capt. Meriwether Lewis and Capt. Wm. Clark (U.S.)	Explore northwestern part of U. S. ascendin__ Missouri River, crossing headwaters of Colum__ bia River and following this river to the Pacifi__
1819	Sir Wm. E. Parry (English)	Discovers Parry Archipelago.
1821	Fabian von Bellinghausen (Russian)	Explores Antarctic and discovers Peter Islan__ and Alexander Island.
1830	Chas. Sturt (English)	Discovers the Murray River and Lake Alexan__ drina, Australia.
1841	Sir James C. Ross (English)	Commands expedition to Victoria Land an__ Volcanoes Erebus and Terror.
1849-1873	David Livingstone (Scotch)	Discovers Lakes Ngami, Shirwa, Nyasa an__ Victoria Falls.
1850	Sir R. M'Clure (Irish)	Discovers Northwest Passage.
1876	H. M. Stanley (Welsh)	Discovers Lakes Albert and Edward.
1881-1884	Lieut. A. W. Greely (U.S.)	Explores Grinnell Land and N.E. coast of Green__ land. Expedition reaches 83° 24½′ N. 1882.
1909	Robert E. Peary (U.S.)	Reaches North Pole with Henson and fou__ Eskimos on April 6.
1909	Sir Ernest H. Shackleton (English)	Reaches lat. 88° 23′ S.; ascends Mt. Erebus__ organizes party which determines the locatio__ of the South Magnetic Pole.
1911	Capt. Roald Amundsen (Norwegian)	Discovers the South Pole on December 18.
1911-1914	Sir Douglas Mawson (British)	Explores Antarctic and discovers King Georg__ V Land.
1914	Col. Theodore Roosevelt (U.S.)	Discovers Roosevelt River, S. America.
1927	C. H. Karius (Australian)	Crosses central New Guinea for first time.
1929	Comm. R. E. Byrd (U.S.)	Makes 1600 mile airplane flight to the Sout__ Pole and back, from base at "Little America"
1933-1935	Rear Adm. R. E. Byrd (U.S.)	Discovers Roosevelt Island on Second Antarcti__ Expedition.
1939-1941	Rear Adm. R. E. Byrd (U.S.)	Explores coast of Palmer Peninsula and adja__ cent islands.
1946-1948	Comm. Finn Ronne (U.S.)	Explores area south of Weddell Sea, proving__ existence of undivided Antarctic continent.
1948	R.C.A.F. fliers (Canadian)	Discover 85 mile long Prince Charles Island i__ Canadian Arctic.
1951	Maj. Franz Risquez (Venezuela)	Discovers source of Orinoco River.
1953	Col. John Hunt (leader) (British)	Commands expedition making first successfu__ ascent of Mt. Everest; summit reached b__ Edmond Hillary and Tensing Norkay.
1958	Cmdr. William R. Anderson (U.S.)	First underwater traverse of the Arctic b__ atomic submarine; reaches North Pole unde__ polar ice pack.
1960	Jacques Piccard (director) (French)	Exploration in bathyscaph reaches record ocea__ depth in Mariana Trench.
1961	Yuri A. Gagarin (U.S.S.R.)	First manned space flight.
1969	Neil A. Armstrong, Edwin E. Aldrin, Michael Collins (U.S.)	First landing of man on the moon.
1973	Charles Conrad, Joseph P. Kerwin, Paul J. Weitz (U.S.)	First manned flight to a space station.